MICROBIOLOGY LABORATORY MANUAL

PRINCIPLES AND APPLICATIONS

SECOND EDITION

STEPHEN A. NORRELL
UNIVERSITY OF ALASKA

KAREN E. MESSLEY
ROCK VALLEY COLLEGE

Prentice Hall

Upper Saddle River, NJ 07458

Editor-in-Chief: Sheri Snavely
Executive Editor: Gary Carlson
Project Manager: Crissy Dudonis
Executive Managing Editor: Kathleen Schiaparelli
Assistant Managing Editor: Dinah Thong
Production Editor: Elizabeth Klug
Supplement Cover Management/Design: Paul Gourhan
Manufacturing Buyer: Ilene Kahn
Cover Photograph: Chronis Jones/Getty Images

© 2003 by Pearson Education, Inc.
Pearson Education, Inc.
Upper Saddle River, NJ 07458

Printed in the United States of America

10 9 8 7 6 5 4 3 2 1

ISBN 0-13-010029-3

Pearson Education Ltd., *London*
Pearson Education Australia Pty. Ltd., *Sydney*
Pearson Education Singapore, Pte. Ltd.
Pearson Education North Asia Ltd., *Hong Kong*
Pearson Education Canada, Inc., *Toronto*
Pearson Educación de Mexico, S.A. de C.V.
Pearson Education—Japan, *Tokyo*
Pearson Education Malaysia, Pte. Ltd.
Pearson Education, *Upper Saddle River, New Jersey*

TABLE OF CONTENTS

PREFACE

ABOUT THE MANUAL

This manual has been written for the Introductory Microbiology student enrolled in either a general or an allied health program. It is intended for those students who take microbiology during their freshman or sophomore year and who, typically, have had only one year (or less) of chemistry. Its aim is to familiarize the student with the basic procedures and equipment used in the microbiology laboratory—including diagnostic microbiology, clinical sample and sampling, serological procedures, antibiotic sensitivity testing and so on.

The manual has several important features. Each exercise has been written to be reasonably self-contained. It features explanations of the salient points being demonstrated or tested. These explanations are meant to supplement and complement, not replace, the assigned text. The exercises have been prepared in a manner that will provide flexibility in their application, dependent on the time frame of the laboratory. This will also enable the exercise to be divided up for performance by individual members of the laboratory group.

DOING YOUR PART

Each exercise is designed to teach a new technique or illustrate a principle of microbial growth. The exercises are arranged in the following sequence:

- Background material discussing the theory and experimental design underlying the procedures used in the exercise;
- Objectives of the laboratory exercise;
- Materials needed for the laboratory;
- Procedures outlining the specific instructions for completing the laboratory exercise; and
- Laboratory Report Form.

As you proceed through the manual, you may notice that the laboratory exercises get progressively more complex and the instructions get somewhat less comprehensive. You are expected to become more competent with each exercise, and you should be able to problem-solve some of the needed procedures for yourself.

Laboratory reports are required for all exercises, and report forms are included for each exercise. They require tabulation of the data acquired in the exercise and include questions based on the experimental procedures and principles demonstrated in the exercise. Many of the questions will require your application of the observed procedures or results to other situations.

LABORATORY SAFETY RULES AND REGULATIONS

Microbiology laboratory exercises are somewhat different from those you might have experienced in most other courses. Microbiology demands the use of live organisms. Most of the experiments in this manual require that you make observations and tests on living and actively growing bacteria. Some of the bacteria are opportunistic pathogens, however, and can be associated with laboratory infections. You must, therefore, use caution to prevent accidental infections. It is always wisest to assume that any culture you are using is pathogenic–then we are less likely to allow ourselves to become careless.

There is also another important reason to develop good techniques for handling material that is potentially infectious. It is very likely that over the course of your career, you will be asked to handle infectious material. Syringes, urine samples, clothing --virtually any item that has been used by a patient is potentially infectious. As a health professional, you are obligated to do all you can to prevent spread of any disease, including those encountered in your professional activity. There are also many instances where the maintenance of sterility is critical (e.g., surgery) and where the introduction of even noninfectious bacteria is a serious circumstance.

During this course, you will be introduced to the safe handling of materials containing microorganisms. This is what is commonly referred to as *aseptic technique*. It is essential for any work with microorganisms. Good laboratory safety practices include the following:

1. Laboratory exercises should be read before the laboratory period and work should be planned.
2. Upon entering the laboratory, place coats, books, and other unnecessary items in the specified location–never on the bench tops.
3. Know the location of all safety equipment, including fire extinguishers and/or blankets and eyewash stations.
4. Wash your hands carefully with liquid soap before and after working in the laboratory. In addition, hands should be washed immediately and thoroughly if contaminated with microorganisms.
5. Carefully wash your work area with disinfectant before you begin your work and after you complete it. This will both limit the spread of dust-borne organisms and destroy any organisms that may have contaminated the work area.
6. Never eat, drink, smoke, or place anything in your mouth in the laboratory.
7. Wear a laboratory coat to protect clothing from contamination or accidental discoloration by staining solutions. In addition, wear shoes at all times in the laboratory and tie back long hair.
8. Be especially careful to develop your aseptic techniques as they are presented to you. Most laboratory infections are the result of poor or lapsed technique.
9. Do not use equipment without instruction.
10. Horseplay will not be tolerated in the laboratory.
11. Do not place contaminated instruments, such as inoculating loops, needles, and pipettes, on the bench tops. Loops and needles should be sterilized by incineration, and pipettes should be disposed of in designated receptacles.
12. Work only with your own body fluids and wastes in exercises requiring saliva, urine, or blood to prevent transmission of disease.
13. Immediately cover spilled cultures or broken culture tubes with paper towels and then saturate them with disinfectant. After 15 minutes of contact, remove the towels and dispose of them in a manner indicated by your instructor.
14. Report accidental cuts or burns to the instructor immediately.
15. Do not touch broken glassware with your hands. Use a broom and dustpan to clean it up, and place the broken glassware in a container as indicated by your instructor.
16. Inoculated media placed in the incubator must be properly labeled and put in the designated area.
17. On completion of the laboratory session, place all cultures and materials in the disposal area as designated by your instructor. All reagents and equipment used must be returned to their proper place.
18. Never remove anything from the laboratory that has not been sterilized without the permission of your instructor. This is especially true for bacterial cultures.

INTRODUCTION TO MICROSCOPY

One of the major tools of microbiology, ever since Antony van Leeuwenhoek first observed his "animacules," has been the light microscope. In this exercise, we will review the compound light microscope, its construction, and its correct use.

BACKGROUND

The microscopes used by van Leeuwenhoek were **simple microscopes**, using only a single lens, which he would grind himself. These microscopes were only able to magnify a total of approximately 300×. Today's light microscopes are **compound microscopes**, which use two or more lenses. The use of multiple lenses and better sources of illumination enable us to significantly improve both magnification and resolution. **Resolution** is the ability to distinguish clearly between two points or objects. Remember, just because an image is enlarged does not mean that we can see it more clearly!

A thorough understanding of the compound light microscope, in terms of its functioning and applications, is very important. The compound light microscope is central to all microbiology laboratories. For the purposes of this discussion, we will consider the microscope as consisting of two systems: the illumination system and the lens system.

THE ILLUMINATION SYSTEM

Proper illumination is essential if we are to view anything clearly with the microscope. Van Leeuwenhoek was limited to the light of the sun or candles to see his specimen. These have been replaced with incandescent or halogen bulbs, which provide readily controlled intensity and color. The focusing system for the light source includes the **iris diaphragm** and the **substage condenser.** The light from the bulb can be adjusted by opening and closing the iris diaphragm. The focusing system for the light source is the condenser. It is designed to deliver a cone of light. The position of the condenser (how far below the stage it is) determines how accurately the diameter of the cone of light matches the opening of the objective lens.

It is important to remember that, as a rule, one must increase the light as one increases the magnification.

It is possible, however, to have too much light illuminating the field of vision. When this occurs, resolution and contrast are lost, making it very difficult to see the slide preparation.

THE LENS SYSTEM

The dual lens system found in a compound microscope consists of the **objective lens**, located closer to the sample, and the **ocular lens**, located closer to your eye. Most microscopes used in the microbiology laboratory have three objective lenses located on a rotating nose piece. These are the low power (10×), high dry (40–45×), and oil-immersion (97–100×) objectives. The ocular lens of most microscopes magnifies 10×. The total magnification of the microscope is determined by multiplying the magnification of the ocular and that of the specific objective being used.

Objective magnification × ocular magnification = total magnification

The maximum usable magnification of a microscope is not limited by the achievable magnification of the lenses, but rather by the **resolving power** of the microscope. Resolving power is the ability to distinguish two points as separate and distinct. **Magnification**, in contrast, simply measures how many times the lens systems of the microscope increase the apparent size. To understand the differences between resolving power and magnification, consider what would happen if you used grainy film to make a highly enlarged photograph—the result would be a large, blurry picture. A microscope with high magnification but poor resolving power produces a large, blurry image—one that is of little value. An ideal microscope is one that has reasonable magnification and good resolution—but if a choice has to be made between the two, it is always in favor of resolution.

The resolving power of a microscope depends upon both the wavelength of light that passes through the system and the numeric aperture (NA), which describes the light-gathering ability of the lens system. Some light is lost as it passes from the illumination system to the lens system owing to refraction, which occurs as the light rays travel from glass to air. The greater the amount of light lost due to refraction, the lower its resolving power. The

to refraction, the lower its resolving power. The refraction of light can be minimized, and therefore the resolving power of the microscope can be maximized, by the use of immersion oil. Immersion oil has approximately the same refractive index as glass. This prevents the refraction of light as it leaves the glass slide, and it gives us a clearer image—better resolution —of the magnified object.

FOCUSING

Modern compound microscopes focus on the subject by moving either the lenses or the stage in relationship to one another. On some microscopes the stage moves up and down; on others the nosepiece moves. In either case there is always a **coarse adjustment knob** and a **fine adjustment knob.** Adjustment is complete when the object is in sharp focus. A competent microbiologist will alternately adjust the light source and the focus to achieve maximum resolution. The various parts of the microscope, illustrated in Figure 1.1, are summarized in Table 1.1. Be sure to locate each of these parts on your microscope.

ROUTINE CARE OF THE MICROSCOPE

A compound light microscope is a precision instrument and requires careful handling. The following rules should be considered mandatory:

Carrying the microscope. The microscope is *never carried with one hand or by the stage.* When you transport the microscope from its cabinet to your work area, one hand supports the microscope from below the base while the other hand grasps the body of the microscope.

Placement of the microscope. Place your microscope securely on your laboratory bench. Do not tilt the microscope unless instructed to do so. Keep the microscope away from the edge of the lab bench.

Peeking and poking. Perhaps the greatest enemy of a microscope is dust and dirt. Every time the lens or any other part of the microscope is removed, dust can enter the light path. You must resist any temptation to look inside the microscope because when you do, the inner surfaces of lenses and prisims can become dirty and dusty. If dust or dirt gets into the ocular or objective lenses, it is very difficult to clean. Do not attempt to do so yourself, but ask your instructor for help. If you have a problem, ask your instructor for assistance—*do not try to fix the problem yourself.*

Cleaning of lenses. *Be sure to clean lenses before and after you use the microscope. Always use lens*

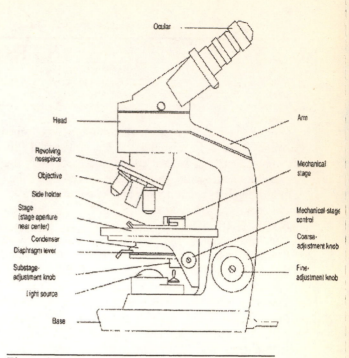

Figure 1.1   The parts of the microscope.

paper to clean lenses. Each lens—the ocular, objective, and condenser—should be cleaned in a circular motion. Fresh lens paper should be used for each lens, with the oil immersion objective being last. Your instructor will guide you in the use of appropriate lens cleaner when necessary.

Storage. The microscope should always be stored in the microscope cabinet. This provides security and allows storage in a relatively dust-free area. Always use the dust cover when provided.

"Your microscope." You should get into the habit of always using the same microscope. This allows you to become familiar with the instrument and any of its idiosyncrasies. Any skilled technician is more comfortable using his or her own tools.

USE OF THE MICROSCOPE

1. Position the microscope in front of you where you can comfortably look into the ocular lens(es).
2. Turn on the light source and position the low-power objective in the light path.
3. Move the substage condenser all the way up, then back it down about one-quarter inch. Be sure that the iris diaphragm is completely open.
4. Look into the ocular lens(es). You should see a full field of view. It is important that, if using a monocular microscope (one with only one ocular), you keep both eyes open. If you

Table 1.1   The Parts of the Microscope and Their
            Function

Arm	Holds the head and stage
Base	Supports microscope
Condenser-adjustment knob	Raises and lowers condenser
Course-adjustment knob	Rapidly brings images into focus
Fine-adjustment knob	Slowly brings image into focus
Head	Holds the ocular(s)
Iris diaphragm lever	Controls amount of light entering the stage aperture
Light source	Illuminates specimen
Mechanical-stage control	Moves slide on the stage
Objectives	Magnifies the image—low power (10×), high dry (40–45×), and oil immersion (97–100×)
Ocular	Magnifies the image, usually 10× or 15×
Revolving nosepiece	Rotates the objectives into position; usually contains low–power, high–dry, and oil–immersion lenses
Slide holder	Secures the slide on the stage
Stage	Holds the slide
Substage condenser	Focuses light on specimen

have a binocular microscope:

a. Adjust the ocular lenses to match the distance between your eyes. As you look into the lenses, pull the lenses apart or push them together until you see a single round field of view. Move them back and forth so you will recognize when they are not in adjustment.

b. The lenses of the binocular head must also be adjusted to compensate for the differences in focusing for each of your eyes. Notice that one lens is adjustable and one is not. When you have a slide on the microscope, you should focus it as sharply as possible as you look through the nonadjustable lens.

c. After the subject is in sharp focus with the nonadjustable lens, turn the adjustable lens until the image is sharp in both lenses. Now, when you use both objective lenses, the image should be in sharp focus for both eyes.

5. Place a prepared slide on the microscope stage and center the slide over the hole in the stage. Be sure that the slide is securely held in place by the mechanical stage or the slide clips.

6. Using the coarse–adjustment knob, position the low–power objective about one-half inch above the slide or as close to the slide as it goes without touching the slide.

7. As you look through the microscope, focus on the specimen. Use the coarse–adjustment knob to bring the specimen into view, then bring the image into sharp focus by using the fine-adjustment knob. **A word of caution:** Most, but not all, modern microscopes are designed so that it is not possible to touch the slide with the lens (they have an **autostop**). You should develop the habit of positioning the lens so that any focusing adjustments will move the lens **away** from the slide.

8. If the light is too bright, reduce its intensity. This may be done in one of three ways: by reducing the voltage to the light by turning a small knob on the light source housing; by moving the substage condenser up and down until the proper light intensity is reached; or by closing the iris diaphragm located on the light source housing by sliding the iris diaphragm lever. It will be necessary to readjust the light intensity when you change lenses.

9. After you are satisfied that you have the low-power objective in clear focus, rotate the high–dry objective over the object. **Another word of caution:** Modern microscopes are **parfocal**. That is, they are designed so that when one lens is in focus (e.g., the low–power objective), the other lenses (e.g., the high–dry objective) will be very near to focus when positioned over the specimen and should only require minor adjustments with the **fine–adjustment knob**. The lenses are also designed so they can be rotated into position without hitting the slide. You need not move the lenses away from the stage to rotate another lens into position. Although most microscopes do have these features, never assume this to be the case. Always watch from the side as you rotate the nosepiece.

10. You will need to focus the lens, but only the fine-adjustment knob should be used. Some adjustments will also need to be done to increase the amount of light. Adjust both the iris diaphragm and the condenser. As before, you should make adjustments to produce the sharpest image possible.

11. After you have carefully focused the high–dry objective, you may proceed to the oil–immersion objective. Remember that you do not need to change focus or move the lenses away from the slide if the microscope is parfocal.

12. Rotate the nosepiece so that the high–dry lens is moved out of position, about midway between the high–dry and the oil–immersion objectives. Place one or two drops of immersion oil on the center of the slide. Continue rotating the nosepiece so that the oil immersion objective is centered over the slide.

The tip of the lens should be in the oil. **A final word of caution***:* The space between the lens and the glass slide decreases with increased magnification. **When using the oil–immersion objective, only use the fine–adjustment knob**. If you cannot focus with the fine-adjustment knob, ask for help.

13. As with the high–dry lens, adjustments will be needed to increase the intensity of light and to increase resolution.

WHEN FINISHED WITH THE MICROSCOPE

1. Using the coarse-adjustment knob, move the stage away from the lens. Remove the slide from the stage. If oil was used, carefully wipe the oil from the slide.
2. Use only lens paper to clean the lenses of the microscope.
3. Using a clean piece of lens paper, wipe the ocular, condenser lens, low and high–dry lenses with lens paper. Finally, remove any oil from the oil–immersion lens.
4. Position the nosepiece of the microscope so the low–power lens is over the center of the stage. Wrap the electric cord securely around the base of the microscope so that the microscope will not be resting on the cord during storage. Cover the microscope with its dust cover, if available.
5. Return the microscope to its storage cabinet, carrying it with both hands. Be careful not to hit the cabinet or shelf with the ocular.

HELPFUL HINTS

1. As you change lenses, move the condenser closer to the objective lens.
2. As you increase magnification, the iris diaphragm will need to be adjusted.
3. As you increase magnification, your working distance between the tip of the objective and the slide will decrease markedly.
4. Proper light adjustment is essential for focusing with the oil–immersion objective. Always adjust your light first, then focus.
5. As you use the microscope, try to remember the relationship between resolution, substage condenser, and light source.

LABORATORY OBJECTIVES

• Utilize all powers of magnification with the compound light microscope.
• Explain how the magnification of a microscope is determined and how the immersion oil allows greater resolution at higher magnification.
• Explain the proper care of your microscope.

• Observe microorganisms using the various degrees of magnification available with your light microscope.

MATERIALS NEEDED FOR THIS LAB

1. Microscope.
2. Prepared slides:
    a. Stained human blood smear
    b. Bacteria, 3 types mixed.
3. Immersion oil.
4. Lens paper.

LABORATORY PROCEDURE

A. USE OF THE LOW- AND HIGH-POWER OBJECTIVES
1. Get your microscope, set it up, and clean all lenses.
2. Place a blood smear slide securely on the stage of your microscope.
3. Follow steps 1 through 8 in Use of the Microscope (pp. 2 and 3). After the slide is in clear focus, and while looking through the ocular, adjust the various light controls. What effect do they have on the viewed image?
4. Move the slide on the stage using the mechanical-stage controls, if available. If your microscope is not equipped with a mechanical stage, carefully move your slide by hand. What happens when you move the slide from left to right? From front to back? Record your observations on the Laboratory Report Form.
5. After observing on low power, rotate the nosepiece of your microscope until the high–dry lens is in position over the slide. Follow steps 9 and 10 in Use of the Microscope (p. 3) to help you properly focus the slide.
6. Use the fine-adjustment knob to focus on the blood cells. Be sure to adjust your light for maximum resolution. How much adjustment did you need to make with the fine-adjustment knob? Which way did you have to turn it? You will find that you will always need to adjust your microscope in the same manner. This is one reason why it is beneficial to always use the same microscope.
7. Move the slide so that you can see the various types of human blood cells. Draw some representative cells on the Laboratory Report Form.
8. Remove the slide from the stage and return to the appropriate place.

B. USE OF THE OIL-IMMERSION OBJECTIVE
1. Obtain a stained bacteria slide that contains all three shapes of bacteria, mixed.
2. Following the above procedures, focus on low power, then on high power. Be sure to carefully adjust your light, as well as the fine–adjustment knob, as you change lenses.
3. Very carefully follow steps 12 and 13 in Use of the Microscope (pp. 3 and 4) as you go from the high–dry to the oil–immersion lens. Before adjusting your focus, adjust your light source so you have a clear and consistent background.
4. On the slide, locate each of the basic bacterial shapes (coccus, bacillus, and spirillum). Draw a few representative cells of each on your Laboratory Report Form.
5. When you have completed looking at your slide, move the stage away from the lens and remove the slide. Carefully blot off any oil from your slide and return it to the appropriate place.
6. Carefully clean the lenses of your microscope with lens paper, being especially careful to remove all oil from the oil–immersion lens. Follow the steps in When Finished With the Microscope (p. 4).

NOTES_____

LABORATORY REPORT FORM
EXERCISE 1
INTRODUCTION TO MICROSCOPY

What is the purpose of this exercise?

A. USE OF THE LOW-POWER AND HIGH-POWER OBJECTIVES

1. When you moved the microscope slide from right to left, which way did the image appear to move?

2. Is your microscope parfocal or nearly so? How can you tell?

3. Draw several representative human blood cells as observed through the high–dry lens.

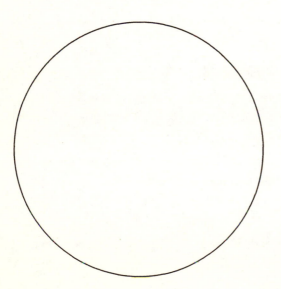

Total Magnification _____

B. USE OF THE OIL–IMMERSION OBJECTIVE

1.   What light adjustment did you need to perform when you went from high–dry to oil–immersion?

2.   Draw cells representing each of the three basic shapes of bacteria.

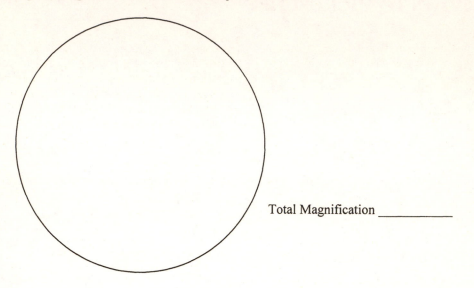

Total Magnification _____

QUESTIONS

   1.   Define:
        a.   Resolution

        b.   Parfocal

   2.   How is the total magnification of a microscope determined?

   3.   What is the purpose of the immersion oil?

   4.   Can you view viruses with your microscope? Why or why not?

   5.   You would like to view the bacterial cells so that they appear larger than possible with your microscope.
        Why can't you simply increase the magnification?

PREPARATION OF SMEARS AND SIMPLE STAINS

BACKGROUND

If we want to learn about individual micro-organisms, we must be able to examine single cells. To do this, we have to stain them and view them through a microscope.

BACTERIAL CELL MORPHOLOGY:
BASIC CELL SHAPES

- **Cocci** are spherical cells. Some cocci, especially those that typically form pairs, may be somewhat flattened on the adjacent sides.
- **Bacilli**, or rods, are cylindrically shaped, straight cells that appear rigid. Typically, but not always, the diameter of the cell is constant throughout its length.
- **Vibrios** are shaped like curved rods; their curvature is always less than a half-circle. As with the bacilli, the diameter is generally constant over the length of the cell.
- **Spirillum** are also curved cells; however, their curvature exceeds that of a half-circle. They are characterized as being spirals rather than gentle curves and may be classified as either spirilla or spirochetes, depending on their exact structure. (See Figure 2.1)

ARRANGEMENTS

An additional morphological characteristic is the clustering of cells following cell division. Although most bacteria separate from each other after dividing is complete, some remain attached. Often this tendency to remain attached produces cell groups characteristic of the species or genus.

The Cocci: Characteristic Cell Groupings

Cocci, like all bacteria, divide by transverse binary fission. Unlike other bacteria, however, some cocci can vary the plane of division and produce clusters of cells that may take the form of long strands, pairs, packets of four or eight, or random clusters. Because these groupings are characteristic for the genus, they do have diagnostic value and are immediately recognized by the experienced microbiologist.

When cells divide along only one plane and remain in pairs, they are referred to as **diplococci**. If they form filaments several cells long, they are considered **streptococci**.

Neisseria and *Branhamella* are genera of cocci that characteristically form diplococci. These two genera typically have flattened adjacent sides, so much so that they resemble kidney beans arranged with the flattened sides adjacent to each other. The genera *Streptococcus, Enterococcus, and Lactococcus* are readily identified by the distinctive chains of cocci present when grown in broth cultures.

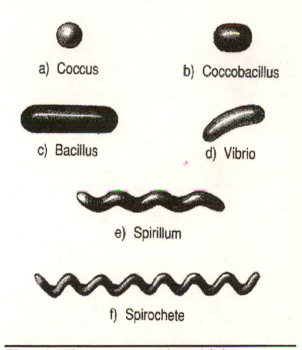

Figure 2.1 The most common bacterial shapes.

The *Staphylococcus* and *Micrococcus* typically form random clusters that have been likened to grape-like clusters or **staphylococci.** This is due to random planes of division during binary fission. The distinction between the arrangements of the pathogenic staphylococci and streptococci is often used as an important part of the diagnostic protocol for the cocci.

Some of the cocci divide in alternating planes and produce packets of four cells—**tetrads**, or eight cells—**sarcinae.** These groupings are formed when the plane of each division is at right angles to the previous one.

The Bacilli: Characteristic Cell Groupings
The bacilli only divide in one plane. This can result in the formation of pairs of cells arranged end-to-end —**diplobacilli**, chains of cells arranged end-to-end—**streptobacilli**, or cells arranged side-by-side—**palisade.**

The diphtheroids, which include *Coryne-bacterium,* are a prominent component of the normal flora of the skin and upper respiratory tract. Some members of the genus are pathogenic. Diphtheroids have a distinctive morphology due to their tendency to remain attached to each other after division. They do not, however, simply form streptobacilli, but rather snap together, producing clusters of cells that have been likened to "palisades" or "Chinese characters." In addition, the cells are often thicker at one end and appear to be club-shaped. The combination of palisade or Chinese-character clusters and club-shaped cells is very distinctive. An experienced microbiologist is able to recognize diphtheroids almost immediately.

The Vibrios and Spirillum
The vibrios and spirillum are usually seen as single cells. The major distinctions between them are based on the degree of curvature they exhibit.

The examples given here are not exhaustive. They do, however, represent examples that have significance in clinical microbiology. Figure 2.2 represents these basic arrangements.

STAINING PROCEDURES
A chemical to be used as a stain for biological material must have at least two properties: it should be intensely chromogenic (colored) and it must combine with some cellular component. When biological stains are used correctly, they enable us to visually examine cells with a considerable level of resolution and definition.

Simple stains are used as general, all-purpose stains. They are usually basic dyes (a salt with the color in the positive ion) that will stain the cell membranes of most bacteria (cell membranes are negatively charged). It is usually necessary to expose the cells to the stain for only a short period of time, during which the positively charged stain will bind with the negatively charged cell membrane. The excess stain is then removed by rinsing before examining the cells with a microscope. In Exercises 3 to 6 we will see how selective reactions can be used to differentiate between cell types and/or cell structures.

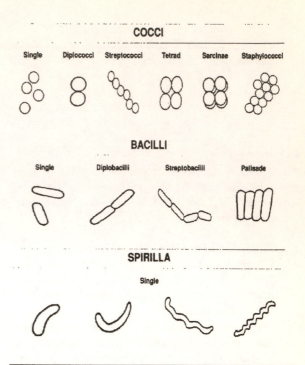

Figure 2.2   Characteristic bacterial arrangements.

PREPARATION AND STAINING OF SMEARS
The preparation of a microscope slide with stained biological matter on it is a two-step process.
1.   **Preparation of smears:** A smear is a slide that has had bacteria placed on it and that has been treated to cause the bacterial cells to adhere to the slide.
2.   **Staining of smears:** Once the bacteria have been adhered to the slide, they must be stained. This involves exposing the cells to the stain, allowing them to react with it for the required period of time, washing the excess stain from the slide, and then drying it.

ASEPTIC TECHNIQUES
Bacteria can be isolated from almost any source. Because they exist virtually everywhere, we must be careful not to contaminate our cultures with bacteria from the air or from instruments. Obviously, neither do we want to contaminate our work area or anything on it (including ourselves!) with the bacterial cultures we are studying.

The precautions used to reduce the chances of contamination are collectively referred to as *aseptic techniques.* If these precautions are properly used, the chance of contamination can be reduced to almost zero.

Your laboratory instructor will demonstrate these precautions as they are needed. They are important and essential for your safety as well as for that of others in the laboratoary. Of course, they are also important to prevent contamination of your culture. These precautions include:

- Flaming of inoculating loops or needles just prior to and immediately following use.
- Prevention of splattering caused by using needles and loops that have not properly cooled after flaming. The most common sources of laboratory contamination and infections are the aerosols created by splattering.
- Flaming the mouths of test tubes and other culture vessels to prevent the inflow of organisms from the air.
- Maintaining the sterility of glassware.
- Proper disposal of contaminated instruments, pipettes, and cultures.
- Proper disinfection of work area before and after use.
- Following designated safety precautions in cases of accidental spills.
- Proper washing of hands with a bactericidal soap both at the beginning and completion of the exercise and any time that spills occur.

LABORATORY OBJECTIVES

- Prepare bacterial smears for microscopic study.
- Stain bacterial smears and observe cell morphology and typical cell arrangements.
- Understand the basic principles underlying staining procedures.

MATERIALS NEEDED FOR THIS LAB

Cultures needed for this lab:
Staphylococcus epidermidis (slant and broth)
Escherichia coli (slant and broth)
Bacillus subtilis (slant)
Corynebacterium xerosis (slant)

A.   SMEAR PREPARATION
   1.   Glass slides: The slides must be carefully cleaned and free of residual oils.

B.   SIMPLE STAINING
   1.   Staining reagents:
         Crystal violet
         Methylene blue
         Safranin
   2.   Glass slides: Use those prepared in part A.

LABORATORY PROCEDURE

Aseptic techniques must be used throughout this exercise. *Follow all instructions carefully and pay special attention to your instructor when these techniques are demonstrated.*

The stains used in this exercise will also stain your fingers and clothing. **Use appropriate care.**

A.   SMEAR PREPARATION
1.   **Preparation of slides**. You will prepare slides as instructed by your laboratory instructor.
   a.   The glass slides must be especially clean. If there is any residual grease on the slide, the smears will be irregularly spread over the glass and it will be difficult to observe bacterial morphology. Wash all slides with soap and water. This should be followed by a water rinse and a final rinse with ethyl alcohol. The slides must be dry prior to their use.
   b.   If you are using slides with a frosted end, you can use a pencil to label the slide; the frosted end will serve as a marker to keep the slide properly oriented (right side up and right to left). If frosted slides are not available, use a marking pen or a wax marking pencil or label your slide.
   c.   Using a marking pen, draw two circles about one-half inch in diameter on each slide, as indicated in Figure 2.3.

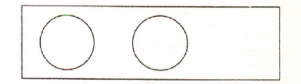

Figure 2.3   Smear preparation

2.   **Preparation of smear from agar slant**. You will need one set of smears for each stain to be used.
   a.   Place a small drop of water in each circle on the slide. Your inoculating loop is handy for applying the proper amount of water.
   b.   Using your flamed inoculating loop, transfer a small amount of culture to the drop. Emulsify the culture in the drop, mixing it thoroughly with the drop of water. This will produce a moderately turbid (cloudy), but not opaque, smear. Spread it evenly throughout the circle. Be sure to flame your loop before *and* after use.
   c.   Repeat until all the slant cultures have been placed on a slide.
   d.   Allow the slides to air–dry.
   e.   Heat–fix the bacteria to the slides by passing each slide through the flame of

your Bunsen burner several times, smear side up. If you do not heat the slide enough, the bacterial cells will be washed off during the staining process. If it is overheated, the cells may become carbonized.

3. **Preparation of smears from broth.** You will need one set of smears for each stain to be used.

   a. Use your flamed inoculating loop to transfer a loopful of culture to your slide. Be sure to mix the broth before removing the loopful. You do NOT need to add water to your slide. Use aseptic technique and remember to flame your loop before and after use.

   b. Spread the drop of broth over the marked area.

   c. If the broth culture is not very turbid, you may need to apply two or three loopfuls to the slide.

   d. Repeat until all the broth cultures have been used.

   e. Allow slides to air–dry.

   f. Heat–fix as described in 2e.

B. STAINING

1. After the slides have completely cooled, place

them on a staining rack located over either the sink or a staining dish or tray.

2. Flood the slides with stain as directed. Make sure each smear is covered completely. Stain each smear for 1 minute. Do not allow the smears to dry.

3. Rinse the excess stain from the slide by holding it under a stream of gently flowing cold water. Let the water flow gently over the smear.

4. Remove excess water from the slide by tapping a long edge of the slide on some paper toweling.

5. Allow the slide to air–dry. Some workers prefer to blot the slide with bibulous paper or paper toweling. If your instructor allows this, place the slide on the paper or toweling, stained side up, and carefully blot it. Do not rub the slide with the paper.

NOTE: Staining procedures, especially simple stains, do not always kill the bacteria, especially not the spore-formers. Blotting the slide may transfer viable bacteria to the paper. Air–drying is the procedure of choice.

---

6. Examine all of the stained smears with the oil-immersion objective. Record your observations on the Laboratory Report Form.

LABORATORY REPORT FORM

EXERCISE 2
PREPARATION OF SMEARS AND SIMPLE STAINS

What is the purpose of this exercise?

SIMPLE STAIN

	Staphylococcus aureus	*Escherichia coli*	*Bacillus subtilis*	*Corynebacterium xerosis*
Appearance (Draw a few representative cells)				
Morphology: Shape				
Arrangement				
Stain Used				
Total Magnification				

QUESTIONS

1. Why is it necessary to heat–fix bacterial smears?

2. What advantages are there to observing unstained bacterial specimens?

WET MOUNTS AND HANGING DROPS

BACKGROUND

Sometimes we want to examine cells while they are still alive. For example, one method for the determination of bacterial motility requires that we watch bacteria move through water or broth. By directly observing bacterial movement, it is possible to distinguish between those organisms which are actively motile and show purposeful movement and those which only exhibit Brownian movement (movement due only to molecular bombardment). Brownian movement appears as irregular, jerky movements. There are two methods by which we can microscopically observe motility:

1. **Wet mounts:** A wet mount is prepared by simply placing a loopful of broth on a slide and then inverting a coverslip onto it so that the drop of broth is between the slide and the coverslip. If the preparation is to be examined over a period of time, petroleum jelly may be used to seal the edges and prevent drying.

2. **Hanging drop:** If a drop of culture is placed directly on a coverslip, it may be carefully placed on a depression slide (there is a circular depressed area in the middle of the slide), allowing the drop of broth to hang in the space of the depression, thus the name "hanging drop." This technique is useful when relatively thick cells, such as protozoans, must be examined.

LABORATORY OBJECTIVES

- Distinguish between motility and Brownian movement.

MATERIALS NEEDED FOR THIS LAB

Broth cultures needed for this lab:

    Staphylococcus epidermidis
    Escherichia coli

A. WET MOUNT
  1. Coverslips
  2. Glass slide
  3. Petroleum jelly
  4. Toothpicks
  5. Broth cultures

B. HANGING DROP
  1. Coverslips
  2. Depression slide
  3. Petroleum jelly
  4. Toothpicks
  5. Broth cultures

LABORATORY PROCEDURE

A. WET MOUNT

1. Carefully transfer a loopful of broth culture to the center of a clean slide. Invert the coverslip over the center of the slide.

2. Lower the coverslip onto the slide so that one edge touches first (see Figure 3.1). The coverslip can then be lowered onto the slide. This procedure will minimize the formation and retention of bubbles in the broth droplet.

3. If you intend to examine the wet mount for more than a few minutes, you should seal the edges with petroleum jelly. Using a toothpick, carefully place a small bead of petroleum jelly around the four sides of the coverslip **before** preparing your wet mount.

4. Invert the coverslip and position it as described above. The petroleum jelly should form a seal when you lower the coverslip onto the slide. A small amount of pressure may be needed to completely seal the edges.

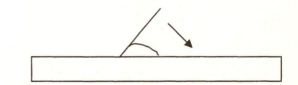

Figure 3.1   Preparation of a wet mount.

5. Examine the wet mount with the high-power objectives. Determine if the bacteria are motile by observing directional movement (as contrasted to simple Brownian movement). *Remember that the bacteria are still viable. Dispose of the wet mount as directed by your instructor.*

6. Complete the Laboratory Report Form.

B. HANGING DROP

1. Place a small bead of petroleum jelly around the outer edges of a coverslip. Place a drop of broth culture in the center of the coverslip.

2. Invert the coverslip over the depression on the slide. Place over the slide so the petroleum jelly seals the coverslip to the slide (see Figure 3.2).
3. Observe under low and high power. Do NOT use the oil immersion objective! Focus on the edge of the drop. Determine if the bacteria are motile by observing directional movement (as contrasted to Brownian movement). *Remember that the bacteria are still viable. Dispose of the*

hanging drop coverslip as directed by your instructor.
4. Complete the Laboratory Report Form.

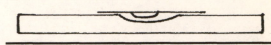

Figure 3.2   Preparation of a hanging drop.

LABORATORY REPORT FORM
EXERCISE 3
WET MOUNTS AND HANGING DROPS

What is the purpose of this exercise?

A. WET MOUNT

	Staphylococcus aureus	*Escherichia coli*
Appearance (Draw a few representative cells)		
Motile or non-motile?		

B. HANGING DROP

	Staphylococcus aureus	*Escherichia coli*
Appearance (Draw a few representative cells)		
Motile or non-motile?		

QUESTIONS

1. Distinguish between true motility and Brownian movement.

THE NEGATIVE STAIN

BACKGROUND

The **negative stain**, as the name implies, stains the background but not the cell itself. Its effect is like that of a photographic negative—everything appears dark except the structure you want to examine. The negative staining procedure is frequently used to examine the capsules of bacteria. It can also be used to aid in the determination of the size and shape of an organism.

The most commonly used procedure is mixing bacteria with India ink or Dorner's nigrosin and examining the resulting suspension as a wet mount (see Exercise 3) or as a very thin smear. India ink is a colloidal suspension of carbon particles. When bacteria are mixed with India ink, the carbon particles cannot penetrate the surface layers of the cell; the cell remains unstained while the background appears dark. Nigrosin is an acidic stain and is repelled by the negatively charged cell membrane, also resulting in the cell remaining unstained against a dark background.

LABORATORY OBJECTIVES

- Understand the role of negative stains in microbiology.
- Gain additional experience in the use of the microscope and in viewing microorganisms.

MATERIALS NEEDED FOR THIS LAB

Cultures needed for this lab:

> *Proteus vulgaris*
> *Pseudomonas aeruginosa*
> *Lactococcus lactis*

WET–MOUNT OR HANGING–DROP
PROCEDURE
1. India ink (high quality) or Dorner's nigrosin
2. Glass slides and coverslips
3. Tubes of approximately 2 ml sterile saline
4. Prepared slides stained with the negative stain.

LABORATORY PROCEDURE

A. WET–MOUNT PROCEDURE
1. If organisms are growing on an agar slant, transfer a loopful of culture into about 2 ml of sterile saline. If they are in broth, you do not need to suspend them in saline.

2. Place one loopful of India ink or Dorner's nigrosin on the slide. Place a loopful of the bacterial suspension next to, *but not touching*, the drop of India ink. Do **NOT** mix the drops at this time.

3. Carefully lower a coverslip over the drops. Allow it to cover both the stain and the suspension of bacteria. When the coverslip is in place, the drops will have mixed in such a way that a stained gradient will be formed.

4. If a sealed wet mount is desired, place the droplets on the coverslip and use the procedure given in Exercise 3.

5. Examine the wet mount with the low-power and high-power objectives. Find the part of the slide that has the optimum contrast for observing both capsules and the refractive bacterial cells.

6. The wet-mount staining procedure does not subject the cells to drying. Any artifacts that occur when cells are dried will be avoided.

7. Record your observations on the Laboratory Report Form.

NOTE: This staining procedure does not kill the bacterial cells. Be sure to soak the slide in a disinfectant following observation.

B. THIN SMEAR
1. Place one loopful of India ink or Dorner's nigrosin near the end of a glass slide. Mix a loopful of bacterial suspension with the stain. Try to keep the droplet as small in diameter as possible.

2. Hold the short edge of another glass slide against the first slide at an angle of about 45 degrees. Beginning near the center of the slide, "back" the tilted slide up until it touches the mixture of stain and bacteria. At this point the mixture should spread out along the back edge of the slide. (See Figure 4.1).

3. Quickly push the tilted side forward until it clears the slide that the smear is on. If the slide was clean and your pushing was smooth, the droplet will have smeared across the slide, producing a thin smear with a feathered edge. Allow the slide to air–dry.

4. Observe the organisms with all three objective lenses, starting with the low–power lens. The best part of the slide is usually near the feathered edge. The cells should be distinctly visible as unstained structures in a smooth-textured dark background.

5. Record your observations on the Laboratory Report Form.

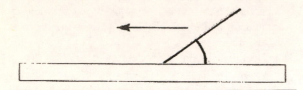

Figure 4.1   Preparation of a negative stain thin smear.

LABORATORY REPORT FORM
EXERCISE 4
THE NEGATIVE STAIN

What is the purpose of this exercise?

ORGANISM			
Appearance (Draw a few representative cells)			
Morphology: Shape			
Arrangement			
Method Used			

QUESTIONS

1.   When might one choose to perform the negative stain?

2.   Can any stain be used for negative staining? Why or why not?

DIFFERENTIAL STAINING PROCEDURES:
GRAM STAIN AND ACID-FAST STAIN

Simple stains, as performed in Exercise 2, stain biological materials indiscriminately. By modifying the staining procedure, using special stains, or adding chemicals, it is possible to differentially stain bacteria. In some cases, only certain *types* of cells will be stained; in other cases, only certain *parts* of cells will be stained. These procedures, which allow us to stain different cells or components different colors, are referred to as **differential staining** procedures.

BACKGROUND

Differential staining, unlike simple staining, requires at least three component factors or steps to make the procedure differential. These include:

- the **primary stain,** used to stain the type of cell or cell component that you want to examine;

- a **mordant,** a chemical that reacts with the primary stain and with the cell or component you want to examine;

- a **selective treatment,** an added step that takes advantage of some special cell characteristic and results in retention of the primary stain by the target cells or cell components. Examples of selective treatment include heating cells or decolorizing with an alcohol solution. *In some procedures both a mordant and a special selective treatment are used;* and

- the **counterstain,** usually a simple stain. It is used to stain everything that was not stained by the primary stain. It is generally a contrasting color.

You will use two differential staining procedures in this exercise—the Gram stain and the acid-fast stain. The Gram stain is the most often used staining procedure in the clinical microbiology laboratory. Other staining procedures will be discussed in Exercise 6.

THE GRAM STAIN

The Gram stain allows rapid distinction between two major groups of bacteria on the basis of the structure and composition of their cell surface membranes and cell walls. It is probably the single most commonly used staining procedure in microbiology. When microbiologists are discussing a bacterial species, the gram reaction is almost always given. For example, they may refer to a "gram-positive bacillus," or a "gram-negative coccus." Virtually all diagnostic protocols used in microbiology begin with the determination of the organism's shape and Gram reaction.

Because of changes that may occur in the bacterial cell wall with age, the Gram-stain technique is most reliable when performed on 24 to 48 hour cultures or on samples directly taken from a patient. When performed on older cultures, gram-positive organisms often appear to be gram-negative.

Primary stain	Gram's crystal violet
Mordant	Gram's iodine
Selective treatment	Acetone–alcohol rinse
Counterstain	Gram's safranin

Gram's crystal violet is applied to a smear and allowed to stain the cells for 1 minute. The crystal violet is then rinsed off the slide and the smear flooded with Gram's iodine. After 1 minute, the smear is rinsed with a mixture of ethanol and acetone. When properly applied, the solvent will remove the crystal violet-iodine complex from gram-negative cells. The counterstain, Gram's safranin, is then used to stain all cells that have not retained the Gram's crystal violet.

Gram-positive cells	Purple
Gram-negative cells	Pink or red

THE ACID–FAST STAIN

The genera *Mycobacterium* and *Nocardia* are two of a very few genera that stain positively with the acid-fast stain. These bacteria have a waxy component in their cell wall that makes the wall almost impermeable to solutes dissolved in water. The wall can be made permeable by heating or may be penetrated by harsh chemicals. The genus *Mycobacterium* includes two important pathogens: *M. tuberculosis* and *M. leprae.* The presence of acid-fast bacteria in sputum or tissue is considered presumptive evidence of either tuberculosis or leprosy. Thus, the acid-fast stain has significance in medical microbiology.

The Kinyoun's procedure is often referred to as **cold staining**. The concentration of carbolfuchsin and phenol has been increased, enabling it to enter the cell without the use of heat. The smear is flooded with Kinyoun carbolfuchsin and allowed to set for 5 to 10 minutes. It is then rinsed with acid-alcohol until the dye no longer flows from the smear. The slide is gently rinsed with water and counterstained with Loeffler's methylene blue. The acid-alcohol will remove all carbolfuchsin that has not been "trapped" inside the acid-fast cells.

Primary stain	Kinyoun's carbolfuchsin
Selective treatment	Acid–alcohol rinse
Counterstain	Methylene blue

An additional method that can be used for the identification of *Mycobacterium* sp. is the application of a fluorescent stain. This is being used increasingly in diagnostic laboratories because of its high specificity. It requires the use of fluorescence microscopy and will not be done during the laboratory sessions.

Acid-fast bacteria	Pink or red
Non-acid-fast bacteria	Blue

LABORATORY OBJECTIVES
- Be able to discuss the components of differential staining procedures and why they differ from simple staining procedures.
- Understand the significance of the Gram-stain procedure and know what components of bacterial cells are responsible for the differential staining reactions.
- Understand the clinical importance of the acid-fast staining procedure and how the staining procedure works.
- Begin to categorize bacteria on the basis of their cellular morphology and staining characteristics.

MATERIALS NEEDED FOR THIS LAB
A. GRAM STAIN
1. Cultures needed (24 to 48 hour)
   Staphylococcus epidermidis
   Escherichia coli
   Moraxella catarhalis
   Corynebacterium xerosis
   Unknown culture
2. Gram-stain reagents:

   Gram's crystal violet
   Gram's iodine
   Ethanol or Acetone-alcohol
   Gram's safranin
3. Prepared Gram stain:
   Staphylococcus epidermidis
   Escherichia coli

B. ACID-FAST STAIN
1. Cultures needed:
   Staphylococcus epidermidis
   Mycobacterium smegmatis (72-hour culture)
2. Acid-fast reagents:
   Kinyoun's carbol fuchsin
   Acid alcohol
   Loeffler's methylene blue
3. Prepared slide of *Mycobacterium tuberculosis*

LABORATORY PROCEDURE
The aseptic techniques required in the previous exercises are also necessary here. Use the same precautions to prevent staining of your fingers, clothing and jewelry.

When differential stains are used, proper controls are needed to ensure that the procedure is working correctly. An easy way to employ such controls is to ensure there is a positive and negative smear on the same slide.

A. GRAM STAIN
1. Prepare smears of bacterial cultures as indicated by your instructor. Heat–fix.
2. Complete the Gram stain. See Figure 5.1.
   a. Flood the smears with Gram's crystal violet.
   b. After 1 minute, remove the crystal violet by rinsing gently with water.
   c. Flood the smear with Gram's iodine and allow it to react for 1 minute.
   d. As you hold the slide over the sink or staining tray, allow the ethanol or acetone alcohol to flow over the smear. As soon as the color stops flowing from the smear, rinse the slide with water. This step, called *decolorization*, is the most critical. It is easy to use too much alcohol, resulting in poorly stained gram-positive cells.
   e. Cover the smears with safranin for 1 minute.
   f. Rinse the stain from the slide, drain the excess water, and air–dry.
   g. Examine all smears with the oil-immersion objective.

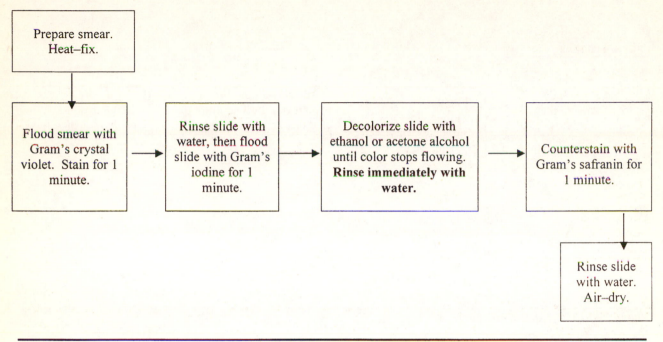

Figure 5.1   Gram–stain procedure.

3.   Staining an unknown organism
      a.   Prepare a slide containing a positive control
           at one end, a negative control at the other
           end, and the unknown in the center   (see
           Figure 5.2).
      b.   Stain as above.
4.   Examine the prepared stains as directed.
5.   Record results on the Laboratory Report Form.

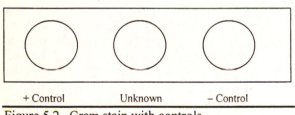

      + Control   Unknown   – Control

Figure 5.2   Gram stain with controls.

B.   ACID-FAST STAIN
1.   Prepare smears of bacterial cultures as indicated
      by your instructor. Heat–fix.
2.   Complete the acid-fast stain (see Figure 5.3).
3.   Flood the smear with Kinyoun's carbolfuchsin
      and allow it to stain for 5 minutes.
4.   Rinse the smear with water to remove excess
      stain.
5.   Flood the smear with acid-alcohol and allow to
      react for 3 minutes. Rinse gently with water.
6.   Apply the acid-alcohol for an additional 1
      minute or until no more red color is removed
      from the smear. Rinse gently with water.
7.   Counterstain with methylene blue for 1 minute;
      rinse, drain, and air–dry.
8.   Examine the prepared slide as directed.
9.   Complete the Laboratory Report Form.

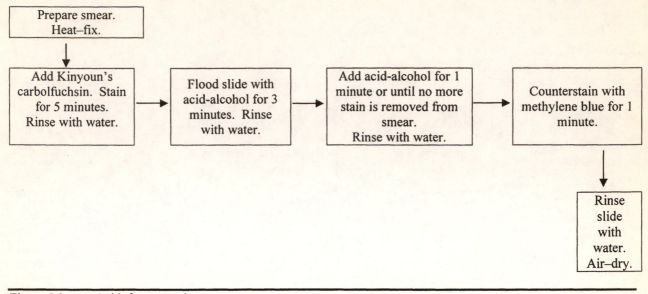

Figure 5.3   Acid–fast procedure.

LABORATORY REPORT FORM

EXERCISE 5
DIFFERENTIAL STAINING PROCEDURES

What is the purpose of this exercise?

A. GRAM STAIN

ORGANISM			
Appearance (Draw a few representative cells)			
Morphology: Shape			
Arrangement			
Gram Reaction			

ORGANISM			
Appearance (Draw a few representative cells)			
Morphology: Shape			
Arrangement			
Gram Reaction			

B. ACID-FAST STAIN

ORGANISM			
Appearance (Draw a few representative cells)			
Morphology: Shape			
Arrangement			
Acid–fast or non-acid–fast			

QUESTIONS

1. A mixed smear of *Staphylococcus epidermidis* and *Escherichia coli* is Gram–stained. Describe how it would appear under the following conditions:
    a. Properly stained smear made from 24-hour cultures.

    b. Properly stained smear made from 7-day–old cultures.

2. Which, if any, step could be omitted in the Gram stain procedure? Why?

3. Why should controls be stained on the same slide as an unknown organism?

4. Name three acid-fast organisms and the disease each causes.

ADDITIONAL STAINING PROCEDURES: METACHROMATIC GRANULES, CAPSULE, ENDOSPORE, AND FLAGELLA STAINS

Four staining techniques are demonstrated in this exercise. These procedures are useful because they stain cellular organelles that would not otherwise be visible.

The structures that are revealed by these procedures are subcellular structures, smaller than the cells that contain them. You will need to pay special attention to getting maximum resolution from your microscope.

BACKGROUND

METACHROMATIC-GRANULE STAIN

Metachromatic granules, or volutin, are storage crystals produced by some bacteria when there is excess phosphate and carbohydrate in their growth medium (a special medium is usually needed). The storage crystals are deposits of cellular phosphate and are large enough to see when stained. Although several kinds of bacteria produce these granules, their production is not common, and those bacteria that do so are relatively distinctive. One of the most commonly encountered species producing the granules is *Corynebacterium diphtheriae,* and the detection of organisms with metachromatic granules is considered presumptive evidence for *C. diphtheriae.*

Albert's Procedure	
Primary stain	Albert's stain
Counterstain	Gram's iodine

Loeffler's Procedure	
Primary stain	Loeffler's methylene blue
Counterstain	None

The metachromatic–granule staining procedure uses a stain that adheres strongly to the phosphate crystals composing the granule. In the Albert's procedure, the smear is covered with the stain and allowed to react for a few minutes. It is then rinsed and counterstained with Gram's iodine. Cells with metachromatic granules will have dark blue to black granules in their cytoplasm, while the cytoplasm appears uniformly light green.

Loeffler's procedure requires simple staining with the methylene blue for several minutes. The granules appear dark blue against a light blue cytoplasm. (See Figure 6.1)

Positive cells	Presence of visible granules within the cytoplasm

CAPSULE STAIN

The **capsule** or **slime layer** is a deposit of slippery, mucoid material that forms around many bacterial cells. Some forms produce capsules that are very small, while others, such as *Klebsiella pneumoniae* and *Streptococcus pneumoniae*, produce prominent capsules. Many studies have linked the capsule with enhanced bacterial virulence, and the capsule has played a fairly prominent role in the history of microbiology (consult our text for information about capsules, bacterial transformation, and the Quelling reaction).

Although it is possible to see the capsule with negatively stained smears, several techniques can be used. In one procedure a negatively stained thin smear is gently heated and then lightly stained. In the Anthony method the capsules are coated with copper sulfate to make them appear light blue.

ENDOSPORE STAIN

Bacterial endospores are highly resistant structures that enable the survival of endospore-forming bacteria when exposed to normally lethal temperatures or from many types of disinfectants and harmful environmental conditions. For example, endospores are not affected by prolonged boiling. As a result of this high resistance to heat, suspensions of bacterial endospores have become the accepted standard for monitoring the effectiveness of most sterilization procedures.

The most commonly encountered endospore-forming bacteria belong to the genera *Bacillus* and *Clostridium,* although at least one coccus is known to make endospores. Because there are very few endospore-forming organisms, the presence of endospores can be diagnostically significant.

Primary stain	Malachite green
Decolorizer	Water
Counterstain	Safranin

The morphology and location of the endospore, relative to the cell that produced it, are quite varied. These variations are used to help classify endospore-producing species. When an endospore-forming species is described, the following characteristics of the endospore are usually noted:

Shape:
    Oval
    Round
Diameter:
    Larger than the cell
    Smaller than the cell
    About the same diameter
Location:
    Central (in the middle)
    Terminal (at the very end)
    Subterminal (in between)

The wall of the endospore is impermeable to water—as well as to anything dissolved in water. However,

the use of a harsh stain allows sufficient dye to enter the endospore so it can be seen.

Bacterial endospore	Green
Vegetative cell	Pink or red

FLAGELLA STAIN

Bacteria that produce **flagella** are motile. A positive test for motility by using the wet-mount or hanging-drop procedures or by the use of motility medium is evidence that the bacterium has flagella. Unfortunately, however, simply knowing whether or not an organism has flagella is often not adequate.

In many protocols for the identification of the gram-negative rods, serological methods are used to test the antigenic nature of the flagella. In other cases it is important to determine the location and number of flagella on a single cell. This information has been found to be consistent for given species and may be used to help classify bacteria into their correct taxons. If they are polarly flagellated, do they have one flagellum at one end, one at both ends, or a tuft of flagella at one or both ends?

Primary stain	Gray's solution "A"
Counter stain	Ziehl-Neelsen carbolfuchsin

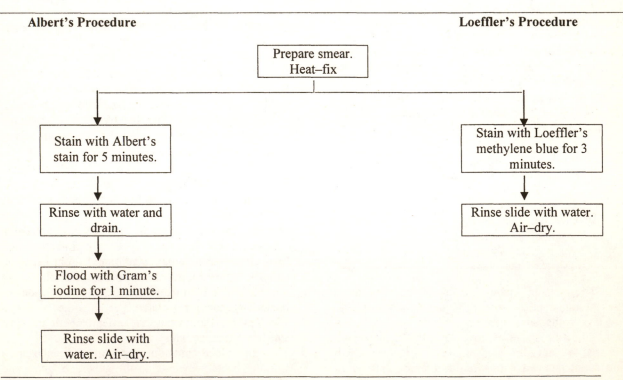

Figure 6.1   Metachromatic-granule stain.

Flagella have diameters considerably smaller than the resolving power of the best optical instrument, even when it is used correctly. The flagella stain makes these organelles visible by coating them with sufficient amounts of mordant (tannic acid and alum) to make them visible. Gray's method requires that (a) the slide is absolutely clean, (b) the cells be handled in an exceptionally gently manner to avoid breaking the flagella, and (c) capsular material be removed before staining.

LABORATORY OBJECTIVES

- Understand the clinical importance of the metachromatic granule staining procedure and know how the staining procedure works.
- Understand the nature of the capsule and its relationship to virulence
- Understand the nature of the bacterial endospore and its relationship to survival of the organism under adverse conditions.
- Understand the nature of bacterial flagella and their role as organelles of locomotion.
- Gain additional experience with the use of the microscope and additional cytological techniques.
- Begin to categorize bacteria on the basis of the cellular morphology and staining characteristics.

MATERIALS NEEDED FOR THIS LAB

A. METACHROMATIC GRANULES
1. Twenty–four–hour cultures of
    Staphylococcus epidermidis
    Escherichia coli
    Corynebacterium xerosis
    Unknown
2. Metachromatic–granule stain reagents
    Alberts stain and Gram's iodine OR
    Loeffler's methylene blue
3. Prepared slide:
    Corynebacterium diphtheriae

B. CAPSULE STAIN
1. Twenty-four–hour cultures of
    Klebsiella pneumoniae
    Streptococcus pneumoniae
2. Capsule stain reagents:
    India ink (high quality) OR
    Dorner's nigrosin
    Crystal violet (1% aqueous)
    Copper sulfate
3. Prepared slides stained with capsule stain

C. ENDOSPORE STAIN
1. Seventy–two–hour cultures of
    Bacillus subtilis
    Bacillus megaterim
    Bacillus cereus
    Clostridium sporogenes
2. Endospore stain reagents:
    Malachite green (7.6% aqueous)
    Gram's safranin
3. Prepared slides stained with endospore stain

D. FLAGELLA STAIN
1. Twenty–four–hour cultures of
    Proteus vulgaris
    Pseudomonas aeruginosa
2. Carefully cleaned slides
3. Sterile saline solution (approx. 2 ml/tube)
4. Flagella stain reagents:
    Gray's solution "A"
    Ziehl's carbolfuchsin

LABORATORY PROCEDURE

A. METACHROMATIC GRANULE STAIN
1. Prepare smears of bacterial cultures as indicated. Heat fix.
2. Complete the metachromatic-granule stain (see Figure 6.1) using the procedure indicated by your instructor.

    Albert's Stain
    a. Flood the smears with Albert's stain for 5 minutes.
    b. Gently rinse the slide with water and drain
    c. Flood the smear with Gram's iodine and allow to react for 1minute.
    d. Rinse, drain, and air–dry.

    Loeffler's Procedure
    a. Flood the smears with Loeffler's methylene blue for 3 minutes.
    b. Rinse, drain, and air–dry.

3. Examine all smears with the oil–immersion objective.
4. Complete the Laboratory Report Form.

B. CAPSULE STAIN (See Figure 6.2)
1. Thin–smear procedure
    a. Prepare a negatively stained thin smear as described in Exercise 4.
    b. Gently heat–fix the smear.

c. Flood the smear with crystal violet for 1 minute.

d. Gently wash the slide with cold water. Air–dry.

e. Examine the slide with all three objectives. The cells should be stained dark blue to purple and should be approximately in the center of the unstained capsule. The background is darkly stained.

f. Record your observations on the Laboratory Report Form.

2. Anthony's Method

a. Prepare a smear as described in Exercise 2. Allow the smear to air dry. Do NOT heat–fix it.

b. Stain the smear with 1% aqueous crystal violet for about 2 minutes.

c. Gently wash the slide with a 20% solution of copper sulfate. Drain. Blot dry.

d. Examine the slide with all three objectives. The cells should appear dark blue or purple, with light blue capsules.

e. Record your observations on the Laboratory Report Form.

C. ENDOSPORE STAIN PROCEDURE
    (See Figure 6.3)

1. Seventy-two hour cultures of *Bacillus* and/or *Clostridium* should be used.

2. Prepare a smear of the cultures in the usual manner. Air–dry and heat–fix.

3. Flood the slide with malachite green. Stain for 10 minutes.

4. Gently rinse slide with water.

5. Counterstain with safranin for 1 minute.

6. Examine the slides with all three objectives. The spores will stain green and the cytoplasm will be pink.

7. Record your observations on the Laboratory Report Form.

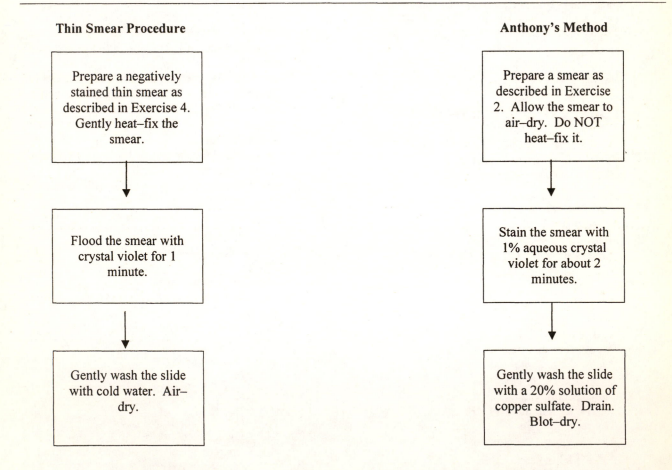

Thin Smear Procedure

Prepare a negatively stained thin smear as described in Exercise 4. Gently heat–fix the smear.

↓

Flood the smear with crystal violet for 1 minute.

↓

Gently wash the slide with cold water. Air–dry.

Anthony's Method

Prepare a smear as described in Exercise 2. Allow the smear to air–dry. Do NOT heat–fix it.

↓

Stain the smear with 1% aqueous crystal violet for about 2 minutes.

↓

Gently wash the slide with a 20% solution of copper sulfate. Drain. Blot–dry.

Figure 6.2   Capsule stain procedure.

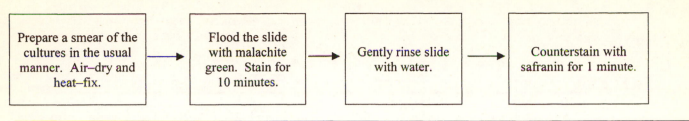

| Prepare a smear of the cultures in the usual manner. Air–dry and heat–fix. | → | Flood the slide with malachite green. Stain for 10 minutes. | → | Gently rinse slide with water. | → | Counterstain with safranin for 1 minute. |

Figure 6.3   Endospore stain procedure.

D. FLAGELLA STAIN PROCEDURE (See Figure 6.4)

1. Sixteen to twenty-four hour broth cultures produce excellent results and may be directly applied to the slide. If it is necessary to use an agar culture, transfer enough loopfuls of culture to produce a distinctly turbid suspension in saline and allow the cells to swim around for about 30 minutes. This will cause the bacteria to lose most of their capsular material.

2. Carefully clean one glass slide for each of the bacteria to be stained. A paste of nonabrasive cleanser can be rubbed on the slide and rinsed in flowing warm water. Either air–dry or wipe with a fresh Kimwipe tissue. After the slide has been cleaned, do not touch the part of the slide that will hold the smear.

3. Place one drop of bacterial suspension near the edge of the slide. Tilt the slide and allow the droplet to flow over the surface. Rock the slide to spread out the smear. Do not use your needle or loop to spread the droplet.

4. Allow the smear to air–dry.

5. Flood the smear with filtered Gray's solution "A." Allow the cells to stain for 8 minutes. Some experimentation may be needed to determine the best staining time—between 6 and 10 minutes usually works well.

6. Gently rinse the slide with distilled water. Do not squirt the water directly on the smear, but let the water flow over it until the stain is removed.

7. Cover the smear with a small piece of filter or blotting paper. Flood the smear with Ziehl's carbolfuchsin for 3 minutes.

8. Remove the paper by lifting (not sliding) it off the smear. Rinse gently with distilled water. Air–dry.

9. Examine the slide with all three objective lenses. You may need to search several sections of the slide to find good examples of the flagellated cells. The cells and their flagella will stain red.

10. Record your observations on the Laboratory Report Form.

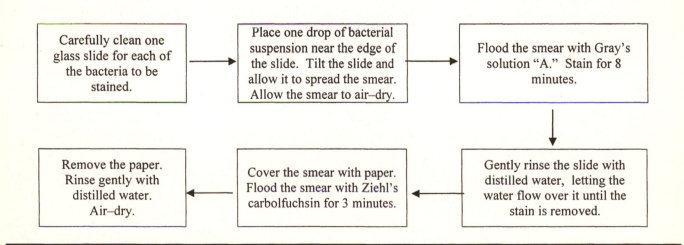

| Carefully clean one glass slide for each of the bacteria to be stained. | → | Place one drop of bacterial suspension near the edge of the slide. Tilt the slide and allow it to spread the smear. Allow the smear to air–dry. | → | Flood the smear with Gray's solution "A." Stain for 8 minutes. |

| Remove the paper. Rinse gently with distilled water. Air–dry. | ← | Cover the smear with paper. Flood the smear with Ziehl's carbolfuchsin for 3 minutes. | ← | Gently rinse the slide with distilled water, letting the water flow over it until the stain is removed. |

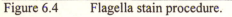

Figure 6.4   Flagella stain procedure.

NOTES

LABORATORY REPORT FORM

EXERCISE 6
ADDITIONAL DIFFERENTIAL STAINING PROCEDURES

What is the purpose of this exercise?

A. METACHROMATIC GRANULES STAIN

Which procedure did you use?

ORGANISM			
Appearance (Draw a few representative cells)			
Morphology: Shape			
Arrangement			
Granules Present or Absent			

B. CAPSULE STAIN

ORGANISM			
Appearance (Draw a few representative cells)			
Morphology: Shape			
Arrangement			
Capsule Present or Absent			

C.   ENDOSPORE STAIN

ORGANISM			
Appearance (Draw a few representative cells)			
Morphology: Shape			
Arrangement			
Endospores Present or Absent			

D.   FLAGELLA STAIN

ORGANISM			
Appearance (Draw a few representative cells)			
Morphology: Shape			
Arrangement			
Flagella Present or Absent			

QUESTIONS

1.   What is the significance of a positive stain for metachromatic granules?

2.   How can the presence of a capsule contribute to an organism's virulence?

3.   What advantage is it for *Clostridium* to form endospores?

4.   List three diseases caused by endospore-forming organisms.

   a.

   b.

   c.

5.   It is possible to observe endospores in simple stained cells. Why then does the endospore stain have to be performed?

6.   What arrangements of flagella can be observed on a flagella-stained cell?

UBIQUITY OF MICROORGANISMS

BACKGROUND

We are constantly hearing of microbes being present—on our body (Wash your hands!), in the water (Don't drink the water in the stream/lake!), on our food (Cook that hamburger until it's well done!), in the air (Be sure to cover your mouth when you sneeze!), and on the ground (Wash that before you put it into your mouth!). It sounds like microorganisms are everywhere!

Something that is always present is said to be **ubiquitous** or omnipresent. Today we are going to try to prove (or disprove) that statement.

Following incubation of the culture, you will be describing the type of colonies you have on your plate. Be sure to compare your plate to see who has similar–appearing colonies. What was their source? Can you form a hypothesis to explain the presence of similar organisms in the varied locations?

LABORATORY OBJECTIVES

- Determine the ubiquity of microorganisms.
- Observe the various cultural characteristics of organisms.
- Examine growth from various sources for cultural similarity.

MATERIALS NEEDED FOR THIS LAB

1. One TSA plate per student
2. Sterile swabs
3. Sterile water tubes

LABORATORY PROCEDURE

1. Choose an object to culture. Use your imagination! The only restriction is that we do NOT want to culture our body with the exception of the skin's surface. Everyone should select a different source for his or her culture. Possible sources:
   a. Any surface in the laboratory, bathroom, or drinking fountain (don't forget such areas as the toilet bowl or faucets)
   b. Air (leave the plate open for 15 to 30 minutes)
   c. Outside (soil, insect, body of water)
   d. Items that we come in contact with or handle (doorknob, computer keyboard, the back of a watch, inside a ring, makeup brush, comb, money)
   e. The surface of your body (touch or kiss the plate, place a hair on the plate)
2. Incubate your plate in an inverted (upside down) position. The temperature of incubation should be close to the source of the sample. The incubator is 37°C. (What is that temperature close to?) Plates can also be incubated at room temperature. Incubate the plates for 24 to 48 hours.
3. Examine the plates for growth. Use the diagrams below to help describe the colonies (See Figure 7.1). Record your observations on the Laboratory Report Form.
4. Perform a Gram stain of the isolated colony types. Record your observations on the Laboratory Report Form.

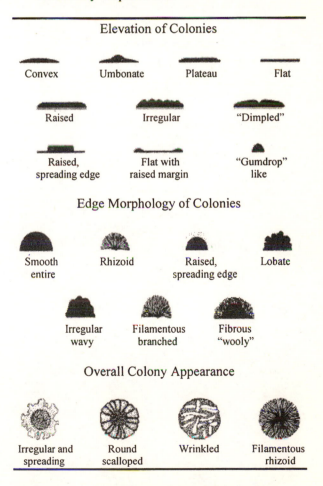

Elevation of Colonies

Convex · Umbonate · Plateau · Flat

Raised · Irregular · "Dimpled"

Raised, spreading edge · Flat with raised margin · "Gumdrop" like

Edge Morphology of Colonies

Smooth entire · Rhizoid · Raised, spreading edge · Lobate

Irregular wavy · Filamentous branched · Fibrous "wooly"

Overall Colony Appearance

Irregular and spreading · Round scalloped · Wrinkled · Filamentous rhizoid

Figure 7.1   Complex colonial morphology.

NOTES_____

LABORATORY REPORT FORM
EXERCISE 7
UBIQUITY OF MICROORGANISMS

What is the purpose of this exercise?

1. What was the source of your culture?

2. Do all colonies on your plate appear the same?

3. Complete the following:

	Colony Type 1	Colony Type 2	Colony Type 3
Appearance (Draw a representative colony)			
Colony color			
Surface smooth or rough appearing?			
Topography			
Edge morphology			
Gram stain			

4. What other sources of cultures had similar–appearing colonies?

5. What hypothesis can you form to explain the similarity of colonies from varied sources?

QUESTIONS

1. Did the class results prove that microorganisms are ubiquitous? Why or why not?

2. Why might some of the microorganisms in the environment been unable to grow on your plates?

SURVEY OF MICROORGANISMS

Microorganisms can be divided into two major groups dependent upon their cell structure—the **prokaryotes** and the **eukaryotes.** Prokaryotic organisms, those which lack a nucleus and other membrane-bound internal organelles, would include the bacteria (both the eubacteria and the archaeobacters) and cyanobacteria. Fungi, protozoa, algae, and the multicellular parasites would be examples of eukaryotic microorganisms. They are all characterized by having a distinct nucleus and membrane-bound organelles. Viruses are not located in either of these groups as they are acellular organisms. Refer to your textbook for further descriptions of these organisms.

In this exercise we will concentrate on examples of the cyanobacteria (prokaryotes) as well as the eukaryotic microorganisms.

BACKGROUND

Most of our laboratory exercises examine the characteristics and identification of bacteria. However, many other organisms are included in microbiology. These include the cyanobacteria, fungi, yeasts, protozoa, algae, and helminths.

The incidence of fungal and parasitic diseases in the United States has shown significant increase over the last several years. Contributing factors appear to be the increased numbers of immunocompromised individuals (those whose immune system is not capable of complete protection), environmental contamination, and our increasingly global society.

KINGDOM MONERA (PROKARYOTAE)

The organisms found in the Kingdom Monera (Prokaryotae) are bacteria and cyanobacteria, as shown in Figure 8.1. The cyanobacteria are photosynthetic, usually unicellular prokaryotic organisms. Some tend to form connected cells that appear as threadlike filaments. Because they are autotrophic (organisms which use carbon dioxide as their major carbon source) they do not pose a health threat to humans. They are, however, extremely important nitrogen-fixing organisms (capable of converting atmospheric nitrogen into ammonia,

which can be used by plants) and are responsible for much of the oxygen produced worldwide.

KINGDOM PROTISTA

Eukaryotic unicellular or colonial organisms are in the Kingdom Protista (see FigureS 8.2A and B). Included are algae and protozoa. Their close relationship can be seen in the dual classification often encountered, such as with the *Euglena*. What characteristics of the *Euglena* would cause its inclusion in the algae? In the protozoa?

ALGAE

Algae are photosynthetic eukaryotes characterized by the presence of membrane–bound chloroplasts and the absence of true roots, stems, or leaves. Many of them exhibit motility by means of flagella or gliding. They are most commonly found in freshwater, with some marine and some soil species seen. Algae are significant primary producers (organisms that convert inorganic carbon as found in carbon dioxide into organic forms, which can be used by other living things), forming the beginning of the food chain. The presence of algae in an aquatic environment can be indicative of its overall quality. When excessive nutrients become present in a freshwater system, it may result in an algal "bloom" or the presence of excessive numbers of algae.

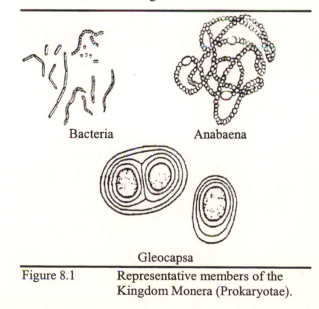

Bacteria   Anabaena

Gleocapsa

Figure 8.1   Representative members of the Kingdom Monera (Prokaryotae).

The microscopic algae, or subkingdom Algae, are often divided into three subgroups—the dinoflagellates (pigmented organisms which are motile by flagella), the diatoms (unicellular algae with silica cell walls), and the green algae.

PROTOZOA

Protozoa are unicellular or colonial, heterotrophic (requiring an organic carbon source) eukaryotes which lack a cell wall. They are usually found in moist and wet areas, and they can be either free-living or parasitic. Classification of the protozoa has traditionally been based on their method of motility—a readily observed trait that may or may not have any evolutionary significance. The groups we will be looking at include the *Sarcomastigophora*, the *Apicomplexa,* and the *Ciliophora*. Human parasitic forms are seen in each of these phyla.

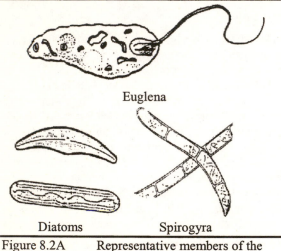

Euglena

Diatoms   Spirogyra

Figure 8.2A   Representative members of the Kingdom Protista—the Algae

The *Sarcomastigophora* include the subphyla *Sarcodina* and the subphyla *Mastigophora*. The *Sarcodina* or those protozoa that move by pseudopodia, include the parasite *Entamoeba histolytica* which causes amoebic dysentery. The *Mastigophora*, or flagellated protozoa, include the parasites *Trichomonas vaginalis* (a sexually transmitted organism), *Giardia lamblia* (the organism largely responsible for protozoan gastroenteritis in day care centers), and *Trypanosoma gambiense* (causative agent of African sleeping sickness).

The *Apicomplexa* contain primarily non-motile protozoa, referred to as the sporozoans, These organisms are often characterized by complex life cycles involving more than one host and are always parasitic. Examples would include the various species

of *Plasmodium* (causative agents of malaria), *Toxoplasma gondii* (a cat parasite that can cause severe congenital defects in humans), and *Cryptosporidium* (responsible for outbreaks of water-borne gastroenteritis).

The ciliated protozoa belong to the group *Ciliophora*. The only human parasite that belongs to this group is *Balantidium coli,* which causes diarrhea.

KINGDOM MYCETAE

Included in the Kingdom Mycetae are both the molds and yeasts. They are unicellular or multicellular, non-photosynthetic eukaryotes, which are surrounded by a chitin or cellulose cell wall and are highly adapted to a variety of environments.

MOLD

The molds are multicellular filamentous fungi composed of individual strands or hyphae organized in a branched meshwork known as the mycelium. They reproduce through the production of asexual and/or sexual spores. The classification of the fungi is based largely upon the type of spore production (see Figure 8.3). The saprophytic molds are important decomposers as they obtain their nutrients from dead organic matter or organic waste. They are also rarely parasitic on healthy plants or animals, instead serving as opportunistic pathogens (an organism that does not usually cause disease but can become pathogenic under special circumstances).

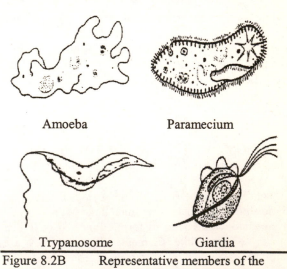

Amoeba   Paramecium

Trypanosome   Giardia

Figure 8.2B   Representative members of the Kingdom Protista —the Protozoa

You will be looking at examples of the *Ascomycota (Aspergillus* and *Penicillium)* and *Zygomycota (Rhizopus).*

YEASTS

The yeasts are nonfilamentous, unicellular fungi which reproduce by budding. Their identification is based upon both morphology and biochemical tests. Yeast are commonly isolated on the surface of plants or amongs the commensal (a relationship between two species where the one species benefits and the other is neither benefited nor harmed) organisms comprising our normal flora. Yeasts are facultative anaerobic organisms, allowing them to grow both in the presence of oxygen, where they undergo aerobic cellular respiration, and in its absence, where they undergo fermentation. This anaerobic fermentation results in the production of alcohol and carbon dioxide and is the basis of the leavening of bread dough and the production of alcoholic beverages.

Yeasts are excellent examples of opportunistic organisms that do not normally cause disease, but can when there is immunosuppression due to illness (AIDS), destruction of other normal flora (antibiotic treatment), or chemical changes in the body (alteration in vaginal secretion with pregnancy).

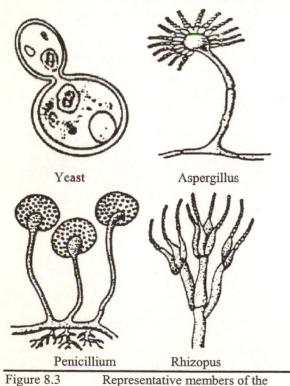

Yeast   Aspergillus

Penicillium   Rhizopus

Figure 8.3   Representative members of the Kingdom Mycetae.

Yeast infections in the forms of thrush or vulvovaginal candidiases are not uncommon under these circumstances.

KINGDOM ANIMALIA—HELMINTHS

The helminths are multicellular eukaryotic organisms that include several phyla of medical and veterinary significance (see Figure 8.4). The Platyhelminths (flatworms) include the flukes and tapeworms. Many of these intestinal parasites have complex life cycles that involve not only man but polluted water, and intermediated hosts such as fish or snails. The Nematodes (roundworms) include the pinworm and hookworms. Many of the roundworms have much simpler life cycles and are present in large numbers in soil, freshwater, and seawater.

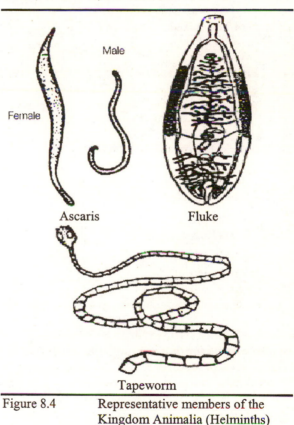

Male

Female

Ascaris   Fluke

Tapeworm

Figure 8.4   Representative members of the Kingdom Animalia (Helminths)

LABORATORY OBJECTIVES

- Distinguish, by kingdom, the eukaryotes observed.
- Name three opportunistic pathogenic molds, the medical problems that they cause in humans, and two beneficial molds.
- Name one pathogenic yeast and the medical problems it causes in humans.
- Identify the distinctive characteristics of each group of protozoans studied.

- Characterize helminths and discuss the reason for their inclusion in microbiology.

MATERIALS NEEDED FOR THIS LAB

A. KINGDOM MONERA (PROKARYOTAE)
1. Living   cultures   or   prepared   slides   of
    Gleocapsa
    Anabaena

B. KINGDOM PROTISTA
1. Living cultures or prepared slides of
    Euglena
    Giardia
    Amoeba
    Plasmodium
    Paramecium
    Trypanosoma
    Algae, mixed
2. Microscope slides
3. Coverslips
4. Pasteur pipettes
5. Methylcellulose
6. Toothpicks

C. KINGDOM MYCETAE
1. Sabouraud agar cultures and/or prepared slides of
    Aspergillus
    Rhizopus
    Penicillium
2. Culture or prepared slide of
    Saccharomyces cerevisiae
3. Prepared slide of *Pneumocystis carinii*
4. Microscope slides
5. Coverslips
6. Pasteur pipettes

D. KINGDOM ANIMALIA
1. Preserved specimen or prepared slides of
    Schistosoma
    Trichinella
    Clonorchis
    Ascaris
    Taenia
    Enterobius

E. MIXED LIVING ORGANISMS
1. Pond water
2. Hay infusion
3. Microscope slides
4. Coverslips
5. Pasteur pipettes
6. Methylcellulose

LABORATORY PROCEDURE

A. KINGDOM MONERA (PROKARYOTAE)
1. Observe the slides of *Gleocapsa* and *Anaebaena*.
2. If living specimens are available, how do they compare with the prepared slides?
3. Record your results on the Laboratory Report Form

B. KINGDOM PROTISTA
1. Prepare wet mounts if living specimens are available.
2. Observe wet mounts using high power. If the organisms are moving too fast, add a drop of methylcellulose to a drop of culture and mix with a toothpick before adding the coverslip.
3. Observe the prepared protozoan and algae slides. How do they compare to the living specimen?
4. Record your observations on the Laboratory Report Form.

C. KINGDOM MYCETAE
1. If Sabouraud agar cultures are available, tease a small amount of growth with a dissecting or inoculating needle. Place in a drop of water on your slide and prepare a wet mount.
2. Observe slides under high power.
3. Observe prepared slides of the fungal specimen. Compare your observations with those of the living specimen.
4. Place a drop of yeast suspension on a clean slide. Add a drop of methylene blue stain. Prepare a wet mount. Observe under high power.
5. Observe a prepared slide of *Pneumocystis carinii*.
6. Record all your observations on the Laboratory Report Form.

D. KINGDOM ANIMALIA
1. Observe the prepared slides and preserved specimen.
2. Record your observations on the Laboratory Report Form.

E. MIXED LIVING ORGANISMS
1. Prepare wet mounts of the pond water and/or hay infusion agar samples. Observe them under high power.
2. Using resources available, attempt to identify the organisms observed. If organisms are moving too fast to view details, add a drop of methylcellulose to your next drop.
3. Record your observations on the Laboratory Report Form.

LABORATORY REPORT FORM

EXERCISE 8
SURVEY OF MICROORGANISMS

What is the purpose of this exercise?

A.   KINGDOM MONERA (PROKARYOTAE)

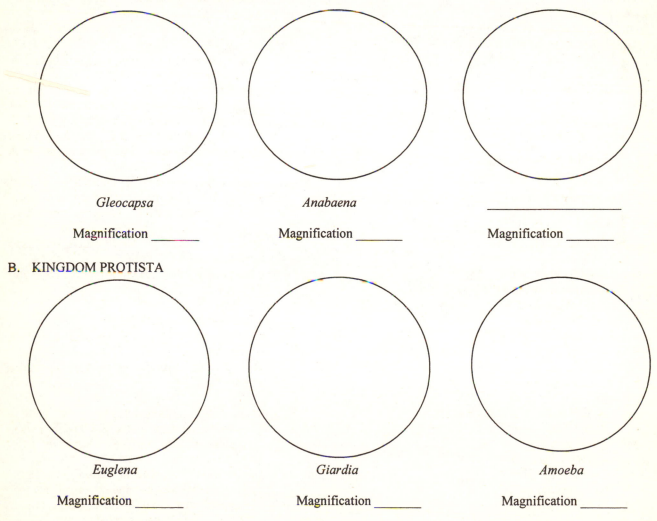

Gleocapsa   *Anabaena*   _____

Magnification _____   Magnification _____   Magnification _____

B.   KINGDOM PROTISTA

Euglena   *Giardia*   *Amoeba*

Magnification _____   Magnification _____   Magnification _____

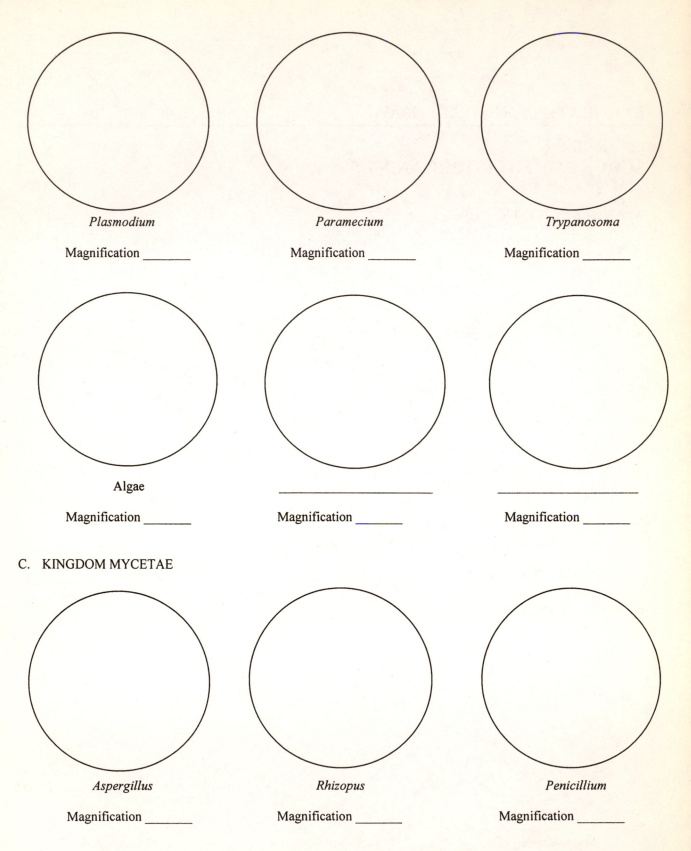

Plasmodium

Magnification _____

Paramecium

Magnification _____

Trypanosoma

Magnification _____

Algae

Magnification _____

_____

Magnification _____

_____

Magnification _____

C. KINGDOM MYCETAE

Aspergillus

Magnification _____

Rhizopus

Magnification _____

Penicillium

Magnification _____

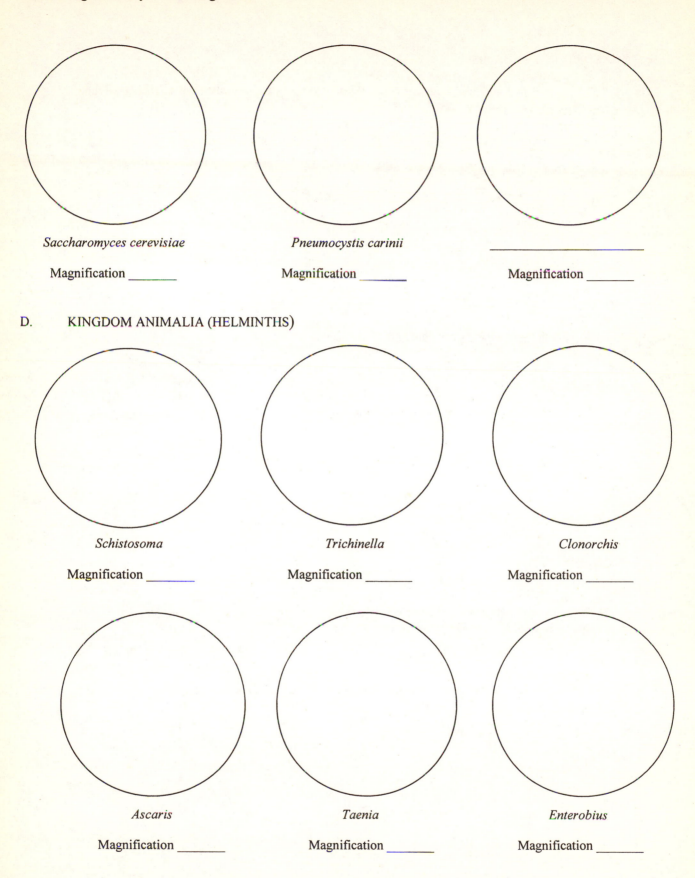

Saccharomyces cerevisiae

Magnification _____

Pneumocystis carinii

Magnification _____

_____

Magnification _____

D.   KINGDOM ANIMALIA (HELMINTHS)

Schistosoma

Magnification _____

Trichinella

Magnification _____

Clonorchis

Magnification _____

Ascaris

Magnification _____

Taenia

Magnification _____

Enterobius

Magnification _____

E.   MIXED LIVING ORGANISMS

Draw any protozoans found in the pond water and/or hay infusion and describe their method(s) of motility.

QUESTIONS

1.   How do the cyanobacteria compare to the bacteria you have previously observed?

2.   What similarities did you observe between the protozoa and the algae? What differences?

3.   Compare mold spores with bacterial endospores.

4.   Give an example of an opportunistic fungi. Are all medically important fungi opportunistic?

5.   Diagram the life cycle of a *Schistosoma*. Propose a public health control plan for the elimination of this organism.

ISOLATION TECHNIQUES: STREAK PLATES

Early in the study of microbiology, Robert Koch recognized that in order to identify the causative agent of disease, it was essential for organisms to be separated from one another and grown in a pure culture. Development of these techniques formed the basis for his Postulates, which are still utilized today.

The procedures that are used in microbiology to accomplish the separation of bacteria into discrete types are referred to as **isolation procedures,** which enable the isolation of bacteria into pure cultures. A **pure culture** is a culture consisting of a single type (species or strain) of bacteria, usually derived from a single cell. Single cells grow into a mass of cells large enough to see. Such a mass of cells is referred to as a **colony**.

Suspensions of cells are spread over the surface of, or within, a nutrient medium in such a way that each cell will grow into a colony that is physically separated from any other colony. Since colonies are assumed to be clones, isolated colonies represent pure cultures.

In this exercise, and the next, we will attempt to isolate pure cultures of bacteria from mixtures. We will also examine the appearance of the colonies and determine the colonial morphology of the organisms studied. It is important to remember that when we isolate a pure culture from a mixed culture, the colonies that grow may contain a bacterium that the microbiologist is seeing for the first time. The recognition of colony types on the basis of their morphology is the critical first step in diagnostic bacteriology.

BACKGROUND

If we are going to be able to grow bacteria, we must provide them, under suitable environmental conditions, with all the nutrients they need. The mixture in which the nutrients are supplied is referred to as the **growth medium** (plural: **media**). The medium also provides the necessary moisture, and it controls the pH of the environment.

Some media are solidified with a colloidal polysaccharide derived from seaweed (red algae), called **agar**. Agar has no nutrient value and cannot be hydrolyzed (broken down to low molecular weight compounds) by most commonly cultured bacteria. It is simply added to the liquid components to solidify the medium. If the medium is used in the liquid form (without the addition of agar), it is called a **broth**. The use of the word **agar** in the name of a medium means that it is a **solid**; the use of the word **broth** means that it is a **liquid** medium.

AGAR

Because agar cannot be hydrolyzed by most bacteria and is of no nutrient value to the bacteria, it can be used as a solidifying agent for bacterial media without changing the components of the medium or without liquefying during the time the bacteria are growing. Agar has several other properties that are also of great importance to the microbiologist, including:

1. **Remains solid at growth temperature.** Agar was discovered as the result of a search for a means of separating bacteria into pure cultures. Gelatin "melts" when the temperature is raised above 25°C and is readily hydrolyzed by many bacteria. When gelatin is used as a solidifying agent it may melt at normal incubation temperatures as well as liquefy owing to bacterial hydrolysis of the gelatin. This results in organisms becoming mixed together; thus they can no longer be studied as pure culture populations. Agar does not liquefy until the temperature is raised above approximately 100°C, and therefore it can be used to solidify culture media without having to be concerned about the mixing of isolated colonies.

2. **Agar is relatively transparent.** Solidified agar is slightly cloudy, but still somewhat transparent. This is an important attribute because it is possible to observe many characteristics of bacterial colonies that would not be apparent were the solidifying agent opaque.

3. **The physical properties of agar.** Agar is a reversible colloid that can change from the sol (liquid) phase to the gel (solid) phase at certain temperatures. The gel to sol (solid to liquid) change occurs at about 100°C, but the reverse change from sol to gel (liquid to solid) occurs at about 45°C. This means that agar can be heated to its liquefaction temperature and then cooled to about 55°C and held in the liquid

state until used. While agar is liquid, additional components can be added to it and, when necessary, bacteria can be mixed in the agar for separation. The agar can then be dispensed into tubes or plates without heat-denaturation of the nutrients or the heat-killing of the bacteria.

THE USE OF AGAR

When agar solidifies, it retains its shape until it is liquefied again. There are several ways that solidified agar is used (see Figure 9.1).

A **slant** is a culture tube (or test tube) that has been maintained in a slanted position while the agar solidified. The tube is filled about 1/3 full with liquefied medium, which is allowed to solidify so that the medium has a substantial slant. The slant provides a good growth surface and is an ideal way of maintaining cultures for study.

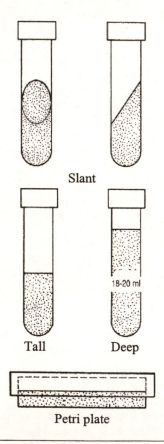

Figure 9.1   Uses of agar.

A **tall,** unlike a slant, is made by allowing the medium to solidify while the culture tube is held in an upright position. The tube is filled between 1/3 and 1/2 full. The media is inoculated using the inoculating needle. The needle, containing the

microorganism, is stabbed straight down into the agar and then pulled out along the same inoculation line.. Stabs are used when reduced oxygen is needed or when it is desirable to observe the effects of growth of bacteria *in* the medium rather than *on* the medium.

A **deep** or **pour** usually contains between 18 and 20 ml of agar. They are usually used to make **pour plates**. Tubes of media are considerably easier to store than are plates, making them easier to handle and manipulate. The medium is liquefied and held in a water bath at about 50–55°C. Components and/or bacteria can be added to it as experimental design requires. Pour plates will be made in Exercise 10.

A **petri plate** is a glass or plastic dish with an overlapping cover. Melted medium is poured into the dish and allowed to solidify. The resulting surface of the medium is used to culture bacteria and to separate them into pure cultures by a procedure known as a **streak plate**. In the streak plate the bacteria are spread *over the surface* of the medium, while in the pour plate, the bacteria are *submerged in* the medium.

ISOLATION PROCEDURES

There are two isolation procedures commonly used in microbiology. The first procedure, the **streak plate**, is one of the most used procedures in any microbiology laboratory. The mixture of bacteria is spread over the surface of a solid nutrient medium so that isolated colonies develop. The second procedure, the **pour plate**, is less commonly used for routine isolation, but is frequently used to estimate the number of bacteria in a sample. It requires that a portion of the sample be mixed with the melted medium, which is then poured into a petri plate and allowed to solidify. If the sample was properly diluted, isolated colonies will develop after incubation. We will perform the pour plate procedure in Exercise 10.

STREAK PLATE PROCEDURE

The successful streak plate will have isolated colonies that can be transferred to fresh media with confidence that the cultures are pure.

1.   A small amount of colony material is picked from the source culture with a sterile inoculating loop or a sterile swab. If the source culture is liquid, a sterile loop is simply dipped into the sample, using proper aseptic technique.

2.   The loop or swab is then streaked over the surface of the nutrient medium is such a way as to spread the sample over the medium. As the loop or swab moves across the surface of the medium, most of the bacteria are deposited at the beginning of the streak. Toward the end, however, the remaining bacteria are placed on the medium so that when they grow, isolated

colonies will develop (see Figure 9.2). It is important that one maintains careful aseptic technique throughout the process or nicely isolated contaminants will be present.

1. Failure to obtain isolated colonies can almost always be traced to one or two common mistakes:

   a. Using *too much inoculum*. If you transfer too much sample to the streak plate, there will not be sufficient surface area to spread out the bacteria. A good rule of thumb is that if you can see more than just a speck of culture material on the end of the needle, you have used too much.

   b. Using *too little surface area*. A single wavy line down the center of the plate will not produce isolated colonies. *Use as much of the surface of the medium as possible*. Place the streaks as close together as possible without overlapping the previously streaked area. The pattern you choose is not nearly as important as whether or not you get isolated colonies.

2. The loop or swab will dig into the media if too much pressure is applied when making the streak. There should be only enough pressure to "steer" the instrument; rely on its weight and balance to provide most of the downward pressure. Practice on the countertop (you do not need to flame the loop this time).

COLONIAL MORPHOLOGY

Bacterial colonial morphology refers to the physical appearance of isolated colonies. Remember, when a colony is grown on an isolation plate, you will often be observing any physical characteristics of the bacterial isolate for the first time. These physical characteristics are often specific for the organism making up the colony and can be used as a means of recognition. Colonial morphology is, however, influenced by the medium and other growth conditions. The colonial morphology of the same bacteria may vary on different media or under differing environmental conditions.

How to Recognize Colonies

Colony recognition is an individual accomplishment. We can illustrate the differences, but you must describe them in terms meaningful to you. For example, if everyone in class was asked to write a description of the laboratory instructor, would you all come up with identical descriptions? Even though your descriptions are different, were any of them wrong? Did any of them fail to accomplish the objective of picking the instructor from a group?

You will be given a list of characteristics that you can use to identify colony types. Whether you use them all or not is up to you—as long as you are able to recognize that colonies do differ and are then able to pick them out of a mixed group of colonies.

Any characteristic that allows you to differentiate and recognize colony types can be used. The list below is meant to be a guide—use whichever characteristics are necessary for you, not your partner, to recognize the colony types.

Colony Physical Characteristics (see Figure 9.3)

1. Colony shape and appearance

Convex	Concave
Flat	Center plateau
Dimpled	"Fried-egg"
shape	

2. Colony consistency

Smooth	Mucoid
Rough	Dry
Fibrous	Granular

3. Colony edge

Spreading	Smooth
Rhizoid	Lobate

4. Colony color
   Color of the colony
   Color of the medium
   Iridescence

5. Changes in medium surface
   Some bacteria dissolve agar and will appear to sink into it.

LABORATORY OBJECTIVES

You will be required to make several streak plates. The streak plates will be made from pure cultures and from a mixed culture containing those same organisms. After incubation of the culture, you will determine the colonial morphology of the pure culture and then attempt to identify it in the mixed culture.

In this exercise you will learn how to isolate mixtures of bacteria into pure cultures and to recognize the specific colony morphology of some bacteria. You should:

- Understand the principles of bacterial isolation.
- Be able to explain the streak plate method.
- Be able to write a description of some bacterial colonies.

MATERIALS NEEDED FOR THIS LAB

1. Twenty-four–hour agar slant cultures of
   Staphylococcus epidermidis
   Pseudomonas aeruginosa
   Serratia marcescens (pigmented strain)
   Bacillus subtilis
2. Twenty-four–hour mixed broth cultures of

Staphylococcus epidermidis
Pseudomonas aeruginosa
Serratia marcescens

3. Nutrient agar plates—1 per pure culture + 1 per mixed culture available. The plates should have been prepared 24 hours ahead of time and incubated. This will allow the plates to dry so that there is no surface moisture to ruin your streak plates. It is also a good quality-control practice to eliminate plates that were accidentally contaminated when they were prepared.

LABORATORY PROCEDURES

1. Obtain and label the nutrient agar plates needed for the streak plates. You will need one for each pure culture and one for each mixed culture.

2. Using aseptic techniques as demonstrated by your instructor, make streak plates of all the cultures. Remember that you should not use too large an inoculum and that you should use as much of the surface of the medium as possible. Avoid digging into the medium. If you make a mistake, repeat the streak plate with a fresh plate, if available.

3. Place all plates in a 37°C incubator in an inverted position (why?). Allow them to incubate for 24 to 48 hours.

4. Following incubation of the cultures, record your observations. Write a description of each colony type and try to identify each organism in the mixed culture plate.

5. If you did not get isolated colonies, examine your streak for possible reasons why. Repeat the experiment until you have obtained isolated colonies if instructed.

6. Record your observations on the Laboratory Report Form.

a) Steps in a three-section streak plate

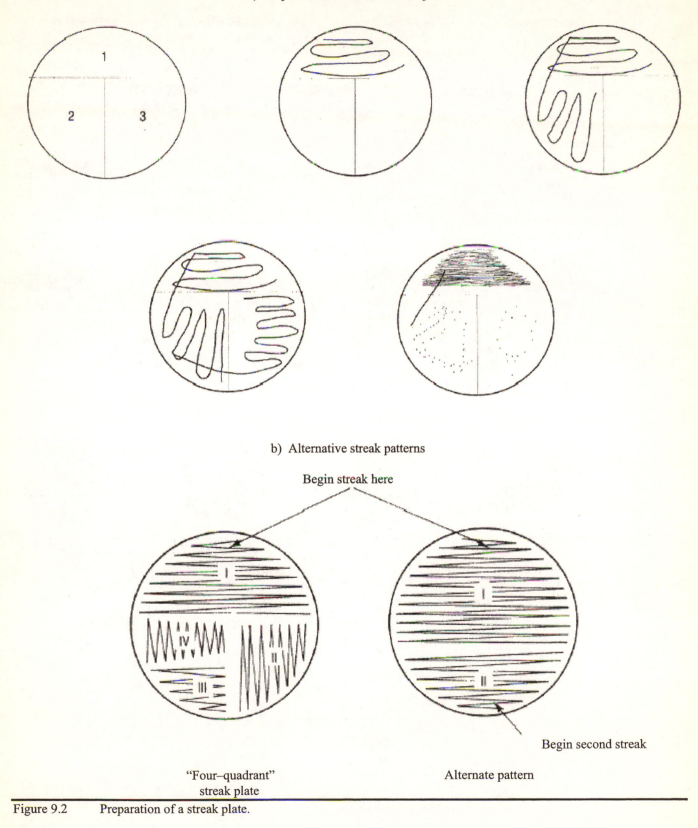

b) Alternative streak patterns

Begin streak here

"Four–quadrant"
streak plate

Alternate pattern

Begin second streak

Figure 9.2   Preparation of a streak plate.

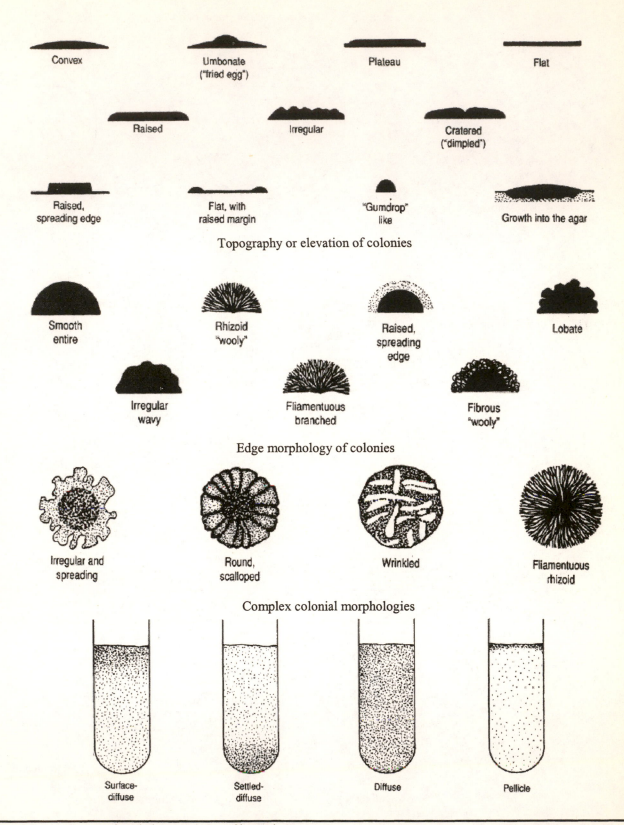

Convex

Umbonate
("fried egg")

Plateau

Flat

Raised

Irregular

Cratered
("dimpled")

Raised,
spreading edge

Flat, with
raised margin

"Gumdrop"
like

Growth into the agar

Topography or elevation of colonies

Smooth
entire

Rhizoid
"wooly"

Raised,
spreading
edge

Lobate

Irregular
wavy

Fliamentuous
branched

Fibrous
"wooly"

Edge morphology of colonies

Irregular and
spreading

Round,
scalloped

Wrinkled

Fliamentuous
rhizoid

Complex colonial morphologies

Surface-
diffuse

Settled-
diffuse

Diffuse

Pellicle

Figure 9.3   Some cultural characteristics of bacteria.

LABORATORY REPORT FORM

EXERCISE 9
ISOLATION TECHNIQUES: STREAK PLATES

What is the purpose of this exercise?

1. Describe the isolated colonies observed on the streak plates made from the pure cultures.

ORGANISM	*Staphylococcus epidermidis*	*Pseudomonas aeruginosa*	*Serratia marcescens*	*Bacillus subtilis*
Colony Appearance (Draw a representative colony)				
Colony Topography				
Colony Consistency				
Colony Edge				
Colony Color				
Changes in Medium				
Amount of Growth (4+, 3+, 2+ 1+)				

2. Sketch the appearance of your streak plate from the mixed culture.

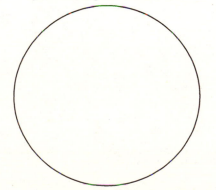

QUESTIONS

1.   How could you determine if contaminants were present on one (or more) of your plates?

2.   Evaluate your streak plates. What was done well? What could you improve on?

ISOLATION TECHNIQUES: POUR PLATES

As we saw in Exercise 9, it is often necessary to isolate bacteria so that we can study a pure culture. One method is the streak plate; another is the pour plate.

BACKGROUND

The pour plate differs from the streak plate in that the sample is mixed with melted medium, which is then poured into a petri plate. After the medium solidifies, the bacteria are held in place by the solidified medium, and isolated colonies develop after appropriate incubation.

1. An aliquot (portion) of a liquid sample is transferred with a sterile loop to liquid medium that is held in a water bath at about 50°C. The medium and sample are then gently mixed (gently enough so that **no bubbles** are formed) and poured into a petri dish. After the medium solidifies, it can be incubated and the isolated colonies observed and selected.

2. If the pour plates are to be used for counting bacteria, they should have between 30 and 300 colonies on them for the counts to be considered valid. More than 300 colonies usually means an overcrowded plate where some colonies are too small to be seen or did not grow at all. What is the reason for the lower limit of 30?

3. Pour plates can be either qualitative or quantitative. In this exercise, we will be performing a qualitative study. In Exercise 22, you will perform a quantitative study.

4. The simplest way of diluting your sample is to transfer a loopful of sample into the melted agar.

In Exercise 9, you determined colony morphology. The same descriptors can be used when discussing surface colonies on pour plates. Those colonies which are embedded within the agar, as in a pour plate, are surrounded by the medium and are limited in their growth by the physical characteristics of the medium. These colonies tend to be lenticular, or foot-ball-shaped, as the colony growth splits the medium, allowing growth only in the "bulge" produced.

LABORATORY OBJECTIVES

The pour plates will be made from either pure cultures or a mixed culture. You should compare the colonies produced by submerged bacteria with those colonies growing on the surface.

In this exercise you will learn how to isolate mixtures of bacteria into pure cultures and to recognize the specific colony morphology of some bacteria. You should:

- Understand the principles of bacterial isolation.
- Be able to explain the pour plate method and contrast it with the streak plate method.
- Be able to describe bacterial colonies.
- Discuss how pour plates can be used for both qualitative and quantitative studies.

MATERIALS NEEDED FOR THIS LAB

1. Three nutrient agar deeps. These are to be melted and tempered (held at 50°C to keep them liquid).
2. **Twenty-four–hour mixed broth culture containing**
   Staphylococcus epidermidis
   Pseudomonas aeruginosa
   Serratia marcescens
3. Three sterile petri plates

LABORATORY PROCEDURE

1. Obtain and label three nutrient agar deeps and three sterile petri plates.
2. Melt the nutrient agar deeps in a boiling water bath. Maintain them in a hot water bath to keep the tubes liquefied at a temperature of approximately 50°–55°C. This process is known as tempering. NOTE: Once you start this procedure, you must continue in a timely manner or the tubes of agar will solidify before you have completed the steps.
3. Take one tube of the tempered agar from the water bath. Make sure that it is still liquid and is still warm to the touch.

4. Using aseptic techniques, add a loopful of the mixed culture to the tube of tempered agar. Mix by rolling the tube between your palms.

5. Remove the second tube of tempered agar and make sure that it is still liquid and warm to the touch. Transfer a loopful of the suspended culture from the first nutrient agar tube to this second tube. Mix the second tube by rolling the tube between your palms.

6. Take the final tube of tempered agar. Transfer a loopful of the suspended culture from the second tube to the third tube. Mix the third tube by rolling the tube between your palms.

7. Pour the contents of each inoculated agar deep into a sterile petri plate. Make sure that the agar is covering the bottom of the plate by gently swirling the plate on the desk. Do not splash.

8. Allow the agar to set completely. Once they have solidified, invert the plates to incubate.

9. Incubate at 37°C for 24 to 48 hours.

10. Observe the number and form of colonies. Use the terms from Exercise 9 to describe colony growth. Record the results on the Laboratory Report Form.

LABORATORY REPORT FORM

EXERCISE 10
ISOLATION TECHNIQUES: POUR PLATE

What is the purpose of this exercise?

RESULTS	Plate 1	Plate 2	Plate 3
Amount of growth (0, +, ++, +++)			
Surface colony appearance			
Subsurface colony appearance			

QUESTIONS

1. How could the pour–plate procedure be used for counting the number of bacteria in a sample?

2. Why do the subsurface colonies look different from the ones growing on the surface of the medium?

3.   What advantages are there to the pour plate procedure? Disadvantages?

OXYGEN REQUIREMENTS OF MICROORGANISMS

A significant number of pathogenic bacteria are anaerobic. In addition to the well-known species of *Clostridium* (*C. tetani, C. botulinum, C. histolyticum,* etc.), there are many anaerobic and microaerophilic bacilli and cocci that are frequently encountered in clinical or environmental samples.

Cultures of the peritoneal cavity and of wounds are perhaps the most likely sources of anaerobic bacteria, although as anaerobic techniques have become more and more reliable, anaerobic bacteria are increasingly being found to be involved in pathogenic conditions. The culture of anaerobic bacteria has therefore become increasingly more important.

BACKGROUND

Bacteria can be classified into four groups according to their requirements for molecular oxygen. In this exercise, these categories will be used to describe the species of bacteria being studied and not to describe a particular type of metabolism or reaction. As we will see, bacterial species that use anaerobic metabolic pathways can be called aerobes if they grow in the presence of oxygen (even though they may not use the oxygen).

1. An **aerobic** bacterial species is able to grow in the presence of molecular oxygen. If the species *requires* molecular oxygen for metabolic energy production, it may be referred to as an **obligate aerobe**; it cannot grow unless molecular oxygen is present in the environment. Such bacteria usually do not ferment sugars, but oxidize them by a respiratory pathway.

2. An **anaerobic** bacterial species is one that does not require the presence of free (molecular) oxygen. Oxygen may be toxic to these bacteria, requiring their growth in environments that are free of oxygen. In some instances, very small amounts of oxygen are toxic, and the requirement for an oxygen-free environment is absolute. Such bacterial species are often referred to as **obligate anaerobes.**

3. Bacteria that require small amounts of oxygen are referred to as **microaerophilic**; they grow best in an environment that has about 5% oxygen. They also prefer an elevated carbon dioxide atmosphere. In a way, the difference between the *obligate anaerobe* and the *microaerophilic* is one of degree. Both groups are considered to be sensitive to oxygen, but one is much more so than the other.

4. Bacteria that are able to grow without regard to the oxygen content of the environment are referred to as **facultative** organisms. They are not inhibited by oxygen, nor do they require it for their metabolism. At least two conditions make an organism facultative:
   a. bacteria that are able to use either an aerobic (aerobic respiration) or anaerobic (anaerobic respiration and/or fermentation) energy-producing metabolism. These bacteria switch from one pathway to another, depending upon whether or not oxygen is present in the environment.
   b. bacteria that use anaerobic metabolic (fermentation) pathways, but which are not inhibited by oxygen. These bacteria will grow in the presence of oxygen, but do not use it. They are sometimes referred to as **aerotolerant** organisms.

For the purpose of this exercise, species of bacteria are defined according to the following:
Aerobe: A bacterial species that grows in the presence of metabolic (free) oxygen.
Anaerobe: A bacterial species that cannot grow in the presence of molecular oxygen. It can be assumed that oxygen is toxic to the cell. These bacteria grow better in an oxygen-reduced environment.
Microaerophile: A bacterial species that grows poorly (but does grow) in the presence of oxygen.
Facultative: A bacterial species that grows well under both aerobic and anaerobic conditions.
Obligate: The term *obligate* may be used as an adjective to imply an absolute requirement for the environmental condition (as "obligate aerobe" or "obligate anaerobe"). When used in this way, the adjectives *obligate* and *facultative* are mutually exclusive.

THE CULTURING OF ANAEROBIC BACTERIA

The techniques used for the cultivation of anaerobes depend upon the removal of free oxygen from the environment. The term *environment* includes both the medium and the atmosphere above the medium. Strict obligate anaerobes require that virtually all molecular oxygen be removed from the medium and from the surrounding atmosphere. Remember that oxygen diffuses easily and that any oxygen remaining in the atmosphere will rapidly dissolve in the medium and diffuse throughout. Strict anaerobiosis is difficult to obtain without special media and equipment.

We will examine two of the techniques commonly used in laboratories to allow the growth of obligate anaerobes as well as one method commonly used to support the growth of microaerophilics.

Thioglycollate Medium

Thioglycollate medium is a nutrient medium that contains thioglycolic acid. This compound is a very strong reducing agent that reacts quickly with any oxygen that might diffuse into the medium. If the thioglycolic acid has been saturated with oxygen, the medium is no longer useful for the cultivation of anaerobes.

Thioglycollate medium is typically used as a semi-solid tube medium and is inoculated with a single stab into the center of the deep. It may, however, be used as a broth or may be solidified with additional agar and used in plates. The use of a small amount of agar to make it semi-solid has the advantage of reducing the diffusion rate of oxygen.

Some commercially available formulas for thioglycollate medium include an indicator dye that is colored in the presence of oxygen. The dye appears colorless when reduced. The use of such a dye has the obvious advantage of providing a convenient means of monitoring the *anaerobiosis*, or absence of oxygen, in the medium.

When thioglycollate medium is correctly inoculated with a single stab down the center of the deep, the pattern of growth produced is distinct for each type of oxygen-related growth (see Figure 11.1). Obligate aerobic bacteria will grow only in the aerobic zone (colored by dye, near the air/medium surface); obligate anaerobes will grow only in the anaerobic zone (uncolored by dye, in the bottom of the tube); microaerophiles will grow slightly below the surface (primarily in the area of the colored/uncolored interface); and facultative bacteria will grow along the entire length of the stab.

As noted earlier, oxygen readily dissolves in most media, including thioglycollate medium, and will quickly saturate the thioglycolic acid. Oxygen, a gas, is also easily removed from solution by heating. If the colored portion of the tube extends more than one–quarter of the depth of the medium, the medium must be heated to boiling to remove the dissolved

oxygen prior to its inoculation. (Why wouldn't you want to boil it following inoculation?) The colored band will disappear upon heating, indicating that the oxygen has been removed. Many laboratories automatically include this step whenever thioglycollate medium is used. It would be essential if you use thioglycollate medium without indicator.

Anaerobic Chambers or Jars

An anaerobic jar is a container that can be rendered anaerobic and which will remain anaerobic as long as it remains correctly sealed. Anaerobiosis is accomplished by using a compound, in this case hydrogen gas, that reacts with oxygen inside the jar, usually reducing it to water. Once the atmospheric oxygen is depleted, the oxygen in the medium will rapidly diffuse into the atmosphere and, in turn, be reduced. We will use the GasPak anaerobic system as an example. (GasPak is a commercial brand name for a product of Baltimore Biological Laboratories, Inc.). The components of the GasPak system (shown in Figure 11.2) are:

1. **Gas generator.** The hydrogen-gas mixture that is used to reduce the oxygen is generated by chemicals provided in a foil envelope. It is only necessary to cut a corner off the envelope, add water, and place the envelope (with the water inside) into the GasPak jar. The reaction that takes place in the jar produces the gas that will be used to reduce the oxygen. Each GasPak envelope may be used only once.

2. **Indicator.** A disposable indicator is available for use in the GasPak system. It consists of a piece of paper saturated with methylene blue. The methylene blue will turn colorless when the oxygen is removed from the atmosphere inside the jar and the methylene blue is reduced. Although an indicator is not essential for the operation of the anaerobic system, it should be used to ensure that anaerobiosis was attained.

3. **Anaerobic jar.** The container consists of the jar itself, a cover, and a clamp that holds the cover tightly on the lip of the jar. Sometimes a sealant, such as a thin layer of petroleum jelly or stopcock grease, may be used to ensure a good, gastight seal between the jar and cover.

4. **Catalyst.** The catalyst safely increases the rate of reaction between the oxygen and the hydrogen gas mixture that is used to reduce it. It is usually contained in a small, screened container that is attached to the inner surface of the cover, or attached to the outside of the gas generator envelope. If these envelopes are used, additional catalyst in the cover of the jar is not needed.

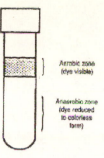

Uninoculated Tube

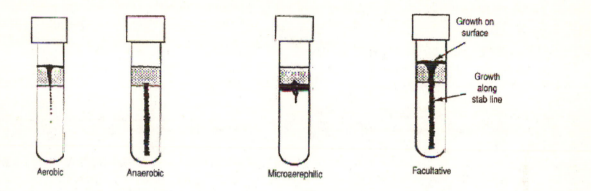

Figure 11.1   Growth in thioglycollate broth.

To use the GasPak system, the following steps should be followed:

1. Obtain a GasPak jar and ensure that the surfaces on the jar and cover are clean and free of deep scratches and defects. Apply a very light coating of petroleum jelly or stopcock grease to the jar's top edge.

2. Check to ensure that a container of catalyst is attached to the inner surface of the cover unless gas generator envelopes with attached catalyst are used.

3. If you are not using a basket that fits inside the jar to contain the petri plates, loosely pack one or two paper towels into the bottom of the anaerobic jar. Moisture will be produced by the catalyzed reaction, and the paper toweling will absorb the moisture.

4. Stack the petri plates into the jar, or the basket that comes with the jar. Be careful that you do not try to use too many plates. When closed, there should be a space of at least one inch between the top petri plate and the catalyst container or cover.

5. Obtain a gas generator envelope and cut off the corner along the dotted line. Place the envelope in the jar, either standing it up between the wall of the jar and the plates or in the holder provided on the basket.

6. Place the indicator strip in the jar in such a way that you will be able to observe it through the side of the jar, or in the holder on the basket. The paper will initially appear blue. It should turn white when anaerobic conditions have been established.

7. Add 10.0 ml of water to the gas generator envelope. The tip of the pipette should not be forced into the envelope.

8. Quickly place the cover on the jar and clamp it in place. The tightening knob on the clamp should be lightly tightened, just enough to make a good seal between the jar and the cover. Do not overtighten the clamp.

9. Place the closed unit in the incubator and incubate for 48 hours. The GasPak jar should not be opened during the incubation period.

CANDLE JAR

In order to grow microaerophiles in a non-reducing medium, a candle jar is used (see Figure 11.3). Inoculated plates or tubes are placed in a jar. A candle is placed in the jar and lit, and the lid of the jar is securely placed on the jar. As the candle continues to burn, the oxygen concentration is diminished and carbon dioxide concentration is increased until there is no longer adequate oxygen to support the combustion of the candle and it is extinguished. The reduced oxygen conditions will be maintained in the jar as long as it remains unopened.

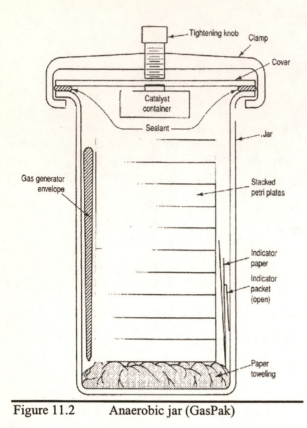

Figure 11.2   Anaerobic jar (GasPak)

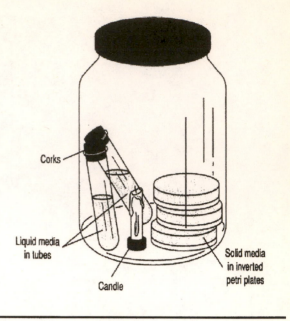

Figure 11.3   Candle Jar

MATERIALS NEEDED FOR THIS LAB

This exercise may be performed individually or in groups as indicated by your instructor.

1.   Twenty-four to 48-hour cultures of the following:
       Clostridium sporogenes
       Pseudomonas aeruginosa
       Staphylococcus epidermidis
       Micrococcus luteus
2.   Five tubes (deeps) of thioglycollate medium. If the dye is visible for more than one–quarter of the depth of the agar deep, the tubes will need to be heated in a boiling water bath to drive off the dissolved oxygen. After you heat the tubes, do not shake them or disturb them while they are cooling.
3.   Three plates of nutrient agar.
4.   GasPak anaerobic jar with catalyst, indicator, and gas generator envelope.
5.   10–ml pipette
6.   Candle jar and candle

LABORATORY PROCEDURE

1.   Obtain one tube of thioglycollate medium for each organism to be tested plus one tube for use as a control.
2.   Label all the tubes and mark the depth of the colored area on the control tube with a marker.

LABORATORY OBJECTIVES

The culture of anaerobic and microaerophilic bacteria requires specialized techniques for the removal of free oxygen. Many pathogenic bacteria are anaerobic or microaerophilic and both their isolation and their identification are essential. To understand the nature of anaerobiosis and to appreciate the special techniques used, you should:

• Understand the meaning of the terms **aerobic**, **anaerobic**, **microaerophilic**, **facultative**, and **obligate** as used in this context.
• Be familiar with simple anaerobic techniques, including the thioglycollate medium and anaerobic jar.
• Be familiar with microaerophilic techniques, including the thioglycollate medium and candle jar.
• Understand the significance of facultative versus obligate organisms in terms of the environmental conditions under which they can grow.

3. Using your sterile inoculating needle, inoculate each tube with a test organism by making a single stab down the center of the deep.

4. Divide three petri plates into quadrants as shown in Figure 11.4. Each quadrant will be inoculated with one of the known organisms. Label one plate "aerobic," one "anaerobic," and one "candle jar."

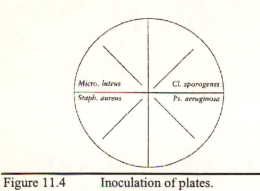

Figure 11.4   Inoculation of plates.

5. Set up the GasPak anaerobic jar according to instructions. Place the plate labeled "anaerobic" in the jar. Follow the directions for creating the anaerobic enviroinment in the jar. Place the jar in the 37°C incubator.

6. Place the petri dish labeled "candle jar" in the candle jar. Place the candle on top of the plates and light it. Securely attach the lid. When the candle is extinguished, place the jar in the 37°C incubator.

7. Place the plate labeled "aerobic" directly in the 37°C incubator.

8. Incubate all cultures for 24 to 48 hours. It is important that the anaerobic and candle jars remain sealed for the entire incubation period.

9. Following incubation, record the growth pattern observed in the thioglycollate medium. Be sure to note the position of any visible growth relative to the colored area.

10. Examine the plates that were incubated aerobically, microaerophilially, and anaerobically. For each organism, compare the amount of growth under each condition, recording growth as 0, +, ++, +++ or ++++. Record your results on the Laboratory Report Form.

NOTES_____

LABORATORY REPORT FORM

EXERCISE 11
OXYGEN REQUIREMENTS OF MICROORGANISMS

What is the purpose of this exercise?

INDICATE YOUR RESULTS FOR THE FOLLOWING:

1. Thioglycollate medium: For each organism, indicate the location of growth. Indicate for each organism whether it appears to be an obligate aerobe, an obligate anaerobe, a facultative, or a microaerophilic organism.

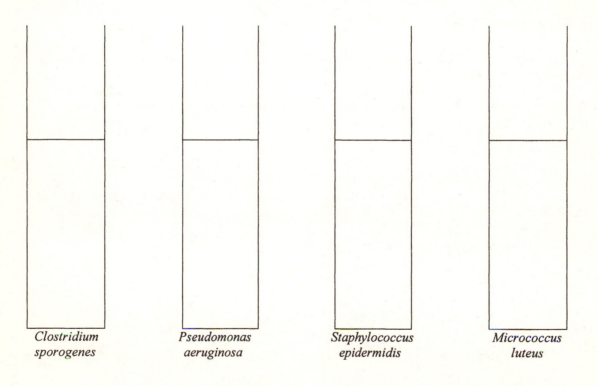

Clostridium sporogenes   *Pseudomonas aeruginosa*   *Staphylococcus epidermidis*   *Micrococcus luteus*

Organism	Oxygen Requirements
Clostridium sporogenes	
Pseudomonas aeruginosa	
Staphylococcus epidermidis	
Micrococcus luteus	

2. Petri plates: Indicate the comparative growth for each organism in the following table. Under *Oxygen Requirement,* indicate whether the organism appears to be an obligate aerobe, obligate anaerobe, facultative, or microaerophilic.

Organism	Aerobic (0, +, ++, +++, ++++)	Anaerobic Jar (0, +, ++, +++, ++++)	Candle Jar (0, +, ++, +++, ++++)	Oxygen Requirements
Clostridium sporogenes				
Pseudomonas aeruginosa				
Staphylococcus epidermidis				
Micrococcus luteus				

3. Do your results for the thioglycollate medium and the plates agree? If not, what explanation might account for the variation?

QUESTIONS

1. List two pathogenic species of *Clostridium* and the disease that each causes.

    a.

    b.

2. List two energy-producing pathways that do not require the utilization of molecular oxygen.

    a.

    b.

3. Humans are considered aerobic organisms. If so, why do we worry about the culturing of anaerobic or microaerophilic organisms?

PHYSICAL GROWTH REQUIREMENTS

Microorganisms are said to be **ubiquitous**, present almost everywhere. This would imply their ability to grow under environmental conditions exhibiting wide variations in temperature, osmotic pressure and pH. To not only survive, but to flourish under diverse conditions, many organisms have developed widely varied adaptive mechanisms. An organism may grow optimally at 25°C, but still be able to survive at 4°C. Other organisms are able to grow in hot springs, a pickle brine, or vinegar. Even so, at some point an organism's adaptive mechanisms for temperature regulation, osmotic pressure or pH can be overcome and vital enzymes or other cell constituents will no longer be able to function properly.

In this exercise we will be examining the optimal temperature, osmotic pressure and pH conditions for several microbial species as well as the varied conditions under which they can survive. Why would it be important to know this information about a given organism?

BACKGROUND
TEMPERATURE
The rate at which chemical reactions take place in a cell is determined by enzyme activity. That temperature at which a cell's enzymes function optimally is referred to as the **optimal growth temperature.** As the temperature of the cell is decreased from its optimum, the rate of enzymatic activity will slow at a rate of approximately 50% for every 10°C drop in temperature. Increased temperature can result in the irreversible denaturing of the enzyme and therefore the cessation of all activity. Figure 12.1 illustrates the relationship between temperature and microbial growth. The **minimum growth temperature** is the lowest temperature at which the species will grow; conversely, the **maximum growth temperature** is the highest temperature at which it can grow. Of course, the **optimum growth temperature** is that at which it grows best.

Bacteria are divided into three major groups based on the temperature at which they grow optimally. Those organisms whose growth range falls between −5°C and 20°C are referred to as **psychrophiles**. Their optimum temperature is around 15°C. These organisms are commonly found in the oceans or in refrigerators where they may be responsible for food spoilage.

Mesophiles are those organisms with optimum growth temperatures between 25°C and 40°C, with many of them growing optimally at 37°C, or human body temperature. Those organisms that comprise the normal flora of humans, as well as human pathogens, are mesophilic. Also included in this group are many of the organisms commonly found in soil. A special group of the mesophiles are the **psychrotrophs**, or those organisms that can also grow at temperatures as low as 0°C. The majority of the organisms we have used or will use in the laboratory are mesophiles.

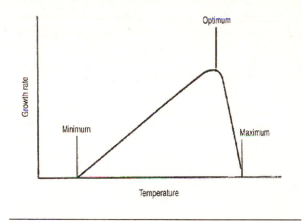

Figure 12.1   Growth curve.

The final group of organisms based on optimum temperatures are the **thermophiles**. These organisms will have a growth range of from 45°C to 65°C although some are able to grow in temperatures greater than 90°C. These organisms are often isolated from hot springs, deserts, or compost piles.

It should be noted that temperature classification (see Figure 12.2) is only a measure of the optimal temperature for an organism's growth and is not reflective of its ability to survive. Endospore-forming organisms, for example, are able to survive in a dormant state over very wide temperature ranges.

The effect of temperature on microbial enzymes can be seen in ways other than growth rates. *Serratia marcescens*, for instance, may only produce prodigiosin, a temperature-dependent red-orange pigment, under specific temperatures.

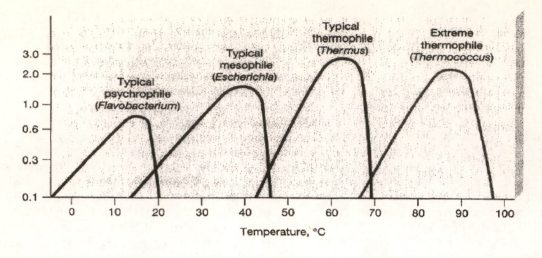

Figure 12.2   Temperature growth ranges for psychrophilic, mesophilic, and thermophilic organisms

In addition to determining the optimum growth temperature for each organism, it is important to know its **thermal death time** and **thermal death point.** The former is the time it takes to kill an organism at a given temperature, whereas the latter is the temperature necessary to kill an organism in 10 minutes. This information is important in the determination of the temperature and time conditions necessary for killing a given organism. You will determine the thermal death time for an organism in Exercise 23.

OSMOTIC PRESSURE

Osmotic pressure is a measure of the force that water exerts on a semipermeable cell membrane due to variations in the concentration of dissolved solutes in the medium or cytoplasm. If the solute concentration is less outside the cell than it is within the cell, water will tend to flow into the cell through the process known as **osmosis**. This movement of water from the point of lower solute concentration to that of higher solute concentration can facilitate the movement of nutrients into the cell and maintain the internal cell pressure.

Those environments exhibiting lower solute concentration than are found within the cell are termed **hypotonic**. Because of the presence of its rigid cell wall, bacterial organisms are not as sensitive to hypotonic solutions as are eukaryotic cells, such as red blood cells, which rapidly undergo hemolysis or rupture when placed in hypotonic solutions.

If the solute concentration is lower inside the cell than it is outside the cell, water will tend to flow out of the cell, resulting in dehydration of the cell as well as possible denaturing of the proteins within the cell. These solutions are **hypertonic** to the cell and often cause cell death. Often hypertonic solutions are used in food preservation. What would be some examples?

Most bacteria grow best in solutions containing solute concentrations close to that of cytoplasm (**isotonic** solutions). Other organisms, however, have had to adapt to growing and reproducing in environments that are hypertonic. Some *Staphy-lococcus* strains will survive in salt concentrations up to 10%. This is important as they inhabit the skin of humans, which, owing to the evaporation of perspiration, can become relatively salty. (How does it feel when you get perspiration in your eye?) Other organisms have become so well adapted to elevated salt concentrations that they cannot survive at lower salt concentrations. These organisms can be isolated from such environments as the ocean (salt concentration approximately 3%), salt lakes (15% to 30% salt concentration), or pickle brines (greater than 15% salt concentration). Organisms requiring a salt concentration of 0.2M NaCl or higher are **halophiles** and can only be grown in higher salt concentrations.

Many of the fungi are **saccharophiles**. What dissolved substance can they withstand in far higher concentrations than can other organisms? How is jelly preserved? Can any microorganism grow on its surface?

VARIATIONS IN pH

We use pH to measure hydrogen ion concentration in a solution. Solutions with increased hydrogen ions are referred to as being acidic and will have a pH value between 0 and 6.9; those with decreased levels of hydrogen ions are basic and will have a pH value between 7.1 and 14. Most bacteria are **neutrophiles**, preferring a pH range of 6.5 to 7.5. As is the case with both temperature and osmotic pressure, we can see the adaptability of microorganisms as we examine the environments in which they can survive.

For example, skin organisms must be able to withstand a pH of around 5 and organisms of the gastrointestinal tract can be exposed to pH levels of 3 or below. Those organisms that can tolerate lower pH values but do not actively grow under those conditions are referred to as being **acidoduric**. Those that not only tolerate but readily grow under those conditions are **acidophiles.** The majority of the acidophilic microorganisms are fungi (fungi have an optimal pH range of 4 to 6). How could you use this information in developing a medium to selectively grow the fungi?

The pH of growth medium is altered by the metabolic activity of organisms growing in it. Sugar fermentation can result in the production of lactic acid. Protein degradation produces amines and ammonia. Either of these reactions could result in the inhibition of bacterial growth due to increased acidity (lactic acid) or alkalinity (ammonia) in the medium. Buffers may be added to media to prevent the resultant change in overall pH of the medium.

LABORATORY OBJECTIVES

Each microorganism must be adapted to live in its natural habitat despite what often appears to be adverse conditions of temperature, osmotic pressure, or pH. To better understand the adaptations that microorganisms have evolved, we must:

- compare and contrast the natural environments of psychrophiles, mesophiles, and thermophiles.
- determine the effect of temperature on bacterial growth.
- define osmotic pressure and explain how it affects a cell.
- compare the effects of isotonic, hypertonic, and hypotonic solutions on microbial cells.
- explain how microbial growth is related to pH.

MATERIALS NEEDED FOR THIS LAB

TEMPERATURE
1. 24-hour cultures of
   Pseudomonas aeruginosa
   Escherichia coli
   Staphylococcus epidermidis
   Bacillus stearothermophilus
   Serratia marcescens
2. Tryptic soy agar (TSA) plates, 8
3. Incubators:
   4°C
   37°C
   55°C

OSMOTIC PRESSURE
1. 24-hour cultures of
   Escherichia coli
   Staphylococcus epidermidis
   Saccharomyces cerevisiae
2. Spore suspension of *Penicillium*
3. TSA plates: 1 each of
   TSA (0% NaCl + 0% sucrose)
   TSA + 5% NaCl
   TSA + 10% NaCl
   TSA + 15% NaCl
   TSA + 10% sucrose
   TSA + 25% sucrose
   TSA + 50% sucrose

pH Levels
1. 24-hour cultures of
   Escherichia coli
   Staphylococcus epidermidis
   Saccharomyces cerevisiae
   Lactobacillus delbruckii subspecies
      bulgaricus
2. Tryptic soy broth (TSB) with pH adjusted to 3.0, 5.0, 7.0, and 9.0 (4 each).

LABORATORY PROCEDURE

TEMPERATURE
1. Divide four TSA plates into quadrants (see Figure 11.4), labeling one section on each plate with
   Pseudomonas aeruginosa
   Escherichia coli
   Staphylococcus epidermidis
   Bacillus stearothermophilus
   Be sure to indicate the incubation temperature on each plate (4°C, room, 37°C, or 55°C).
2. Using a sterile loop, inoculate the center of each section with a short straight line using the appropriate organism.
3. Incubate each plate at the designated temperature for 24 to 48 hours.
4. Label the remaining TSA plates for inoculation with *Serratia marcescens*. Be sure to include the incubation temperature on each plate (4°C, room, 37°C, or 55°C).
5. Using a sterile loop, prepare a streak plate with *Serratia marcescens* on each.
6. Incubate each plate at the designated temperature for 24 to 48 hours.
7. Compare the amount of growth of each organism at the varied temperatures. Record the growth as 0, +, ++, +++, or ++++.

8. Compare the pigment production of *Serratia marcescens,* noting both the color and its intensity at the various temperatures.

OSMOTIC PRESSURE
1. Divide each plate into quadrants. Label each quadrant with one of the organisms.
2. Using your sterile loop, inoculate the center of each section with a short straight line of the appropriate organism.
3. Incubate the plates at 37°C for 24 to 48 hours.
4. Compare the relative amounts of growth of each organism under the varied osmotic conditions. Record the growth as 0, +, ++, +++, or ++++ on the Laboratory Report Form.

5. Incubate the plates at room temperature an additional 3 to 5 days. Record the relative amounts of growth on the Laboratory Report Form.

pH LEVELS
1. Inoculate each organism into one tube of TSB at each of the pH levels.
2. Incubate the tubes for 24 to 48 hours.
3. Resuspend the organisms in each broth. Determine the relative amounts of growth, comparing each organism with itself at varied pH levels. Record the results as 0, +, ++, +++, or ++++ on the Laboratory Report Form.

LABORATORY REPORT FORM

EXERCISE 12
PHYSICAL GROWTH REQUIREMENTS

What is the purpose of this exercise?

A. TEMPERATURE

1. What is the room temperature of the laboratory?

2. Complete the following table, indicating the relative amount of growth at each temperature (++++, +++, ++, +, or 0).

Effect of Temperature on Growth

Organism	4°C	Room Temperature	37°C	55°C
Pseudomonas aeruginosa				
Escherichia coli				
Staphylococcus epidermidis				
Bacillus stearothermophilus				

3. Compare the appearance of *Serratia marcescens* at various temperatures:

	4°C	Room Temperature	37°C	55°C
Amount of Growth				
Appearance (pigment)				

4. Attempt to classify each of the following organisms as psychrophilic, mesophilic, or thermophilic.

Pseudomonas aeruginosa _____

Escherichia coli _____

Staphylococcus epidermidis _____

Bacillus stearothermophilus _____

B. OSMOTIC PRESSURE

1.   Record the following results, indicating the relative amount of growth at each temperature (++++, +++, ++, +, or 0).

Effect of NaCl Concentration on Microbial Growth

Organism	0% NaCl	5% NaCl	10% NaCl	15% NaCl
Escherichia coli				
Staphylococcus epidermidis				
Saccharomyces cerevisiae				
Penicillium				

Effect of Sucrose Concentration on Microbial Growth

Organism	0% Sucrose	10% Sucrose	25% Sucrose	50% Sucrose
Escherichia coli				
Staphylococcus epidermidis				
Saccharomyces cerevisiae				
Penicillium				

C. pH LEVELS

1.   What is the natural habitat for each organism? What pH level would you expect to be its optimum?

Organism	Natural Habitat	Predicted Optimum pH
Escherichia coli		
Staphylococcus epidermidis		
Saccharomyces cerevisiae		
Lactobacillus delbruckii subspecies *bulgaricus*		

1.   For each of the following organisms, report its relative growth at the various pH levels.

Organism	pH 3.0	pH 5.0	pH 7.0	pH 9.0
Escherichia coli				
Staphylococcus epidermidis				
Saccharomyces cerevisiae				
Lactobacillus delbruckii subspecies *bulgaricus*				

QUESTIONS

1.   An organism is inoculated onto TSA plates and incubated at 4°C, room temperature, 37°C, and 55°C for 48 hours. The following results are noted:

4°C	Room Temperature	37°C	55°C
++	++	++++	0

How would you classify this organism based on temperature growth?

2.   Based on your experimental results, what is the medical significance of endospore-forming organisms?

3.   Why was it necessary to inoculate a plate containing 0% NaCl and 0% sucrose?

4.   What organisms are most likely to be involved in the spoilage of jelly?

SELECTIVE AND DIFFERENTIAL MEDIA

We saw in Exercises 11 and 12 that different organisms have varied physical growth requirements that must be met if we are to successfully culture them. In addition to their physical growth requirements each organism also has specific nutritional requirements. For some species, they will only need to have a surface to grow on, limited inorganic salts, and air. Others are dependent on the provision of complex organic molecules to meet their nutritional needs.

Media can also be utilized to provide us with additional information about a given organism. By adding pH indicators, alternative sources of carbon or nitrogen, or inhibitory chemical agents we can determine the varied nutrients an organism is capable of utilizing, the by-products of metabolism produced, or the organism's ability to withstand adverse chemical agents.

BACKGROUND

Bacteriological media can serve many purposes in the microbiology laboratory—from enabling the collection of specimens, transport of specimen to the laboratory, detection, primary isolation, and identification of microorganisms, and the determination of antimicrobic susceptibility. In this exercise we will concentrate on those media that aid in the primary isolation and identification of microorganisms.

We will only be exploring the utilization of a few of the hundreds of types of media available to the microbiologist. The choice of media will depend upon the source of the organism or specific organisms being screened for. Complete information about the composition and uses of varied media can be found in the *Difco Manual* (Difco Laboratories) and the *BBL Manual* (Becton-Dickinson).

TRANSPORT MEDIA

One of the most critical factors in clinical microbiology is the adequate preservation of the organism from the time it is collected from the patient or environment to the time of culturing in the laboratory. Transport media are non-nutrient, semi-solid medias that inhibit self-destructive enzymatic reactions within the cell. It also is formulated to prevent the lethal effects of oxidation. Examples of transport media include Amies Transport Media and Stuart Transport Media. Transport media will be discussed further in Exercise 31.

GENERAL–PURPOSE MEDIA

Nonselective media used for primary isolation of microorganisms are the general–purpose media. These media will support the growth of a variety of normal body flora, pathogens, and soil microbiota. To encourage the growth of organisms, enrichment factors such as blood, hemoglobin, and growth factors may be added. Examples of general–purpose media include Nutrient Agar, Tryptic Soy Agar, and Blood Agar for bacterial growth and Sabouraud Dextrose Agar for fungi.

SELECTIVE MEDIA

Often, when attempting to isolate organisms, we are not interested in those organisms that comprise the normal flora but instead are looking for the presence of other species not routinely present. This could include screening for the presence of *Escherichia coli* in a body of water (indicative of water contamination with fecal material) or for the presence of mycobacteria in a sputum sample. To facilitate the detection of these organisms among the normal flora, selective media are used.

Selective media are formulated to suppress or inhibit the growth of one group of microorganisms while allowing the growth of others present in a mixed flora. This is usually achieved by the incorporation of bacteriostatic agents or the alteration of the physical or chemical environment of the media. Examples would include the inclusion of crystal violet, which, in concentrations of 1:100,000, is inhibitory to gram-positive organisms but not to gram-negative organisms. Other dyes, antibiotics, sodium chloride, sodium azide, phenylethanol, and bile salts are just a few of the commonly used selective agents. Examples of selective media include Columbia CNA Agar, Phenylethanol Medium and Lowenstein-Jensen Agar.

ENRICHMENT MEDIA

At times, the presence of large numbers of normal flora can overgrow and obscure the presence of pathogens. Enrichment media are designed to suppress the competing normal flora, enabling the growth and further isolation of the pathogen.

Examples include GN Broth for the selective cultivation of *Salmonella* and *Shigella* from stool specimen and Selenite Cystine Broth for the enrichment of *Salmonella* from food and dairy products.

SPECIALIZED ISOLATION MEDIA
Specialized isolation media are formulated to satisfy the needs of a specific microorganism, thereby allowing its more rapid isolation and identification. An example would be Mannitol Salt Agar, which is selective for the pathogenic staphylococci.

DIFFERENTIAL MEDIA
A differential media incorporates certain reagents or chemicals into the media, which results in recognizable reaction following incubation. For example, if a fermentable sugar, such as glucose, is present in the media as well as a pH indicator, it will provide, following incubation, a quick visual distinction between those organisms which are fermenters and those which are non-fermenters.

It is possible for a single medium to be both a general–purpose medium and a differential medium. We have already discussed Blood Agar as a general–purpose medium, capable of supporting the growth of a large number of microorganisms. Certain organisms, however, cause specific visible changes in the media due to their varied production of hemolysins (enzymes that cause the rupture of red blood cells). If the organisms are able to lyse the blood cells, it results in a complete clearing of the agar around the colony—a process referred to as beta-hemolysis. If this is present on a Blood Agar plate inoculated with a throat culture it is highly significant for the presence of the beta-hemolytic streptococci or "strep throat."

Other media have been formulated to be both selective and differential—providing chemical agents, such as dyes, that will inhibit the growth of many organisms and also providing other chemicals, such as fermentable carbohydrates, that will enable the differentiation between those that are growing. An example of this is Eosin–Methylene Blue (EMB) Agar. This medium is commonly used for the detection of fecal organisms in water supplies. A free-standing body of water will contain many different organisms, most of which are derived from soil. The dyes eosin and methylene blue are inhibitory to the growth of those gram-positive cells that are present in the water. Many of the gram-negative organisms that are present, such as *Enterobacter aerogenes*, are found in soil runoff and need to be distinguished from those of possible fecal origin, such as *Escherichia coli*. By incorporating lactose ("milk sugar"), which is not fermented by most soil organisms, it is possible to presumptively determine the identity of *E. coli* based on the production of acid (seen as a green metallic sheen).

LABORATORY OBJECTIVES
- Define all-purpose, selective, and differential media.
- Distinguish between bacteria by utilizing all-purpose, selective, and differential media.
- Demonstrate and judge the advantages of a medium that is both selective and differential.
- Understand the relationship between the growth of an organism and the composition of the medium.
- Learn the advantages of the use of selective, differential, and enriched media in the laboratory growth of microorganisms.

MATERIALS NEEDED FOR THIS LAB
1. Broth cultures of
       Escherichia coli
       Enterobacter aerogenes
       Staphylococcus epidermidis
       Streptococcus faecalis
2. Media, one of each per group
       Tryptic Soy Agar plate
       Phenylethanol Agar plate
       Mannitol Salt Agar plate
       Levine EMB Agar plate
       Hektoen Enteric Agar plate
       Blood Agar plate
       Columbia CNA plate

LABORATORY PROCEDURE
1. Label each plate with the name of the media and your identifying information. Divide each plate into quadrants. Label each section with the name of one of the test organisms (See Figure 11.4).
2. Inoculate each section with the appropriate organism.
3. Incubate all plates at 37°C.
4. Following incubation of cultures, record the results for each of the plates. The media and results are summarized in Table 13.1.
    a. Tryptic soy agar: observe the plates for the presence of growth. What would it mean if there was no growth on one of the sections?
    b. Phenylethanol agar: observe the plate for the presence of growth.
    c. Mannitol salt agar: observe the plate for the presence of growth. For those organisms that grew, record the color of the surrounding medium (pink or yellow).

d. Levine EMB agar: observe the plate for the presence of growth. For those organisms that grew, record the color of the colonies (pink, metallic green, or colorless).

e. Hektoen enteric agar: observe the plate for growth. For those organisms that grew, record the color of the colonies (salmon-pink, green, with or without black centers to colony) and the agar surrounding the colonies (pink zones or no change).

f. Blood agar: observe the plate for the presence of growth. For those organisms that grew, record the hemolysis present as alpha-hemolysis (zone of greening around the growth), beta-hemolysis (zone of clearing around the growth), or gamma-hemolysis (absence of hemolysis).

g. Columbia CNA agar: observe the plate for the presence of growth. For those organisms that grew, record the type of hemolysis observed. Hemolysis is described, as with the blood agar.

Table 13.1   Selective and Differential Media.

Media	Substrate	Selective Agent(s)	Organisms Selected For	Reaction and Description
Tryptic Soy Agar		None	General–purpose media	
Phenylethanol Agar		Phenylethyl alcohol	Gram-positive bacteria	Growth or no growth
Mannitol Salt Agar	Mannitol	7.5% Sodium chloride	*Staphylococcus aureus*	1. Mannitol-fermenter (colonies surrounded by yellow zones) 2. Non-mannitol fermenter (small colonies with no yellow zones)
Levine EMB Agar	Lactose Sucrose	Eosin Y Methylene blue	Gram-negative bacilli	1. Lactose-fermenter (dark purple colonies or colonies with dark centers and transparent colorless borders) 2. Non-lactose or non-sucrose fermenters (colorless colonies
Hektoen Enteric Agar	Lactose Sucrose Salicin Amino acids containing sulfur	Bile salts	Gram-negative bacilli	1. Lactose-fermenter (salmon-pink colonies) 2. Non-lactose-fermenters (green, moist colonies) 3. Salicin-fermenters (pink zones around colonies) 4. Non-salicin–fermenters (no change) 5. H_2S producers (colonies with black centers)
Blood Agar	Hemoglobin	None		1. Alpha–hemolysis (green zones around colonies) 2. Beta–hemolysis (clear zones around colonies) 3. Gamma–hemolysis (no zone around colonies
Columbia CNA Agar	Blood	Colistin Nalidixic acid	Gram-positive bacteria	1. Alpha–hemolysis (green zones around colonies) 2. Beta–hemolysis (clear zones around colonies) 3. Gamma–hemolysis (no zone around colonies

LABORATORY REPORT FORM

EXERCISE 13
SELECTIVE AND DIFFERENTIAL MEDIA

What is the purpose of this exercise?

1. Record your results on the following chart:

Medium	*Escherichia coli*	*Enterobacter aerogenes*	*Staphylococcus epidermidis*	*Streptococcus faecalis*
Tryptic Soy Agar Growth + or				
Phenylethanol Agar Growth + or				
Mannitol Salt Agar Growth + or				
Appearance				
Levine EMB Agar Growth + or				
Appearance				
Blood Agar Hemolysis + or				
Alpha-, Beta -, or Gamma-				
Columbia CNA Growth + or				
Appearance				

2.   Based on your results, which media would you consider to be:

   a.   General Purpose

   b.   Selective

   c.   Differential

   d.   Selective and Differential

QUESTIONS

   1.   Describe the basis for selectivity of EMB agar, Hektoen Enteric agar, and Mannitol Salt agar.

   2.   A urine specimen is sent to the clinical laboratory indicating that it is from a female patient with a possible urinary tract infection. The microbiologist plates it on EMB agar. Why?

   3.   When you go to the doctor's office with pharyngitis (a sore throat), the nurse will often take a swab of your throat and plate it on blood agar. Why?

   4.   You are working in a clinical laboratory. When might you choose to use the following media? (You may find it helpful to read about each in the *Difco Manual*.)

   a.   Phenylethyl Alcohol (Phenoethanol) Medium

   b.   Deoxycholate Medium

PHYSIOLOGICAL CHARACTERISTICS OF BACTERIA: CARBOHYDRATE METABOLISM

The next three exercises cover the physiological or biochemical characteristics of bacteria.

There are at least two reasons for studying the biochemical characteristics of bacteria. First, these characteristics can be used to demonstrate the exceptional metabolic diversity of prokaryotic organisms. The range of metabolic capabilities that bacteria exhibit is very large, explaining in part why these organisms can be found in virtually every environmental habitat.

The second reason is that the biochemical characteristics of bacteria represent additional phenotypic characteristics that can be easily examined. Because individual species' characteristics are genetically determined, it is possible to use them as phenotypic markers. It is then possible to identify unknowns by matching the phenotype of the unknown to that of a known reference organism. A *reference organism* is one considered to be typical of a given genus and species—so much so that it can be used as a reference against which unknown isolates can be compared.

BACKGROUND

The reactions of carbohydrates that are employed for identification are usually degradative catabolic reactions used by the bacteria as part of their energy-producing metabolism. The synthesis of structural and storage carbohydrates is also important, but these reactions are not often used in identification protocols. The reactions you will study fall into two major categories: those which determine *if* certain carbohydrates are used at all, and those which determine *how* they are used. For example, we will test several bacterial species to determine which sugars they are able to use (i.e., glucose, lactose, etc.) and whether or not the sugars can be fermented with the production of acid and gas, or only acid.

A NOTE ABOUT pH INDICATORS

The pH indicators will change color as the pH of the solution they are in changes. The actual color change takes place over a range, the center of which is referred to as the pK. The indicators you will use in this exercise are shown in Table 14.1.

These data indicate that phenol red will be yellow below pH 6.8, that it will begin to turn red at 6.8, and that it will be completely changed by pH 8.4. Between 6.8 and 8.4 it will be various shades of orange, becoming increasingly more red as the pH increases and increasingly more yellow as the pH decreases. Brom thymol blue will change from yellow to green to blue as the pH rises from 6.0 to 7.6.

Table 14.1   Select pH Indicators

Indicator	pK	Range and Color
Methyl red	5.2	4.4–6.0 Red-Yellow
Phenol red	7.9	6.8–8.4 Yellow-Red
Brom thymol blue	7.0	6.0–7.6 Yellow-Blue

EXPERIMENTAL METHODS

To determine an organism's biochemical characteristics, you must use a medium that induces or enhances that characteristic, and you must use some chemical tests to measure the activity. This often involves the use of pH indicators in the medium to detect the production of either acids or bases, but may also depend upon added chemicals that react with the products to give a colored compound.

FERMENTATION OF SUGARS

Sugar-fermentation broths contain the sugar being tested and a pH indicator to detect the production of acid. In this exercise, you will use phenol red sugar broths. If the sugar is fermented, the medium will turn from red to yellow because of the production of acid. The glucose sugar tube will also have a smaller tube inside for the entrapment of gas bubbles, should any gas be produced. This is referred to as a Durham tube. If the bacteria do not use the sugar, they will often use the amino acids in the medium. When amino acids are used, ammonia is produced as a by-product, causing the pH of the medium to rise (the color will turn deeper red). Additionally, if organisms have depleted the carbohydrates, they can then start breaking down amino acids. This can result in a pH change from acid to alkaline. If the sugar is used oxidatively (respiration), there will be little, if any, color change. Sugar fermentation results are reported as *acid* (A), *acid + gas* (AG), *alkaline* (Alk), or *no change* (NC).

HYDROLYSIS OF STARCH

When iodine and starch react, a deep purple color will develop. If a starch agar plate is flooded with an iodine solution, any starch present will react with the iodine, causing a deep purple color to form. If the starch has been hydrolyzed around the colonies, the medium will remain unstained. The enzyme secreted by bacteria that hydrolyze starch is called *amylase*; the test is occasionally referred to as the *amylase test*.

METHYL RED–VOGES–PROSKAUER (MR-VP) TEST

Methyl Red (MR) Test

Methyl red is a pH indicator that changes color at a lower pH (about 4.4) than phenol red. When a few drops are added to a broth containing a sugar, a color change will result if the pH falls below about 4.4, You should remember that this indicator turns red below a pH of 4.4 and yellow at a pH above that point. Unlike the phenol red used in the fermentation broths, a **red color is positive** for the methyl red test. This test is used in most identification schemes for the gram-negative rods.

Voges-Proskauer (VP) Test

This is a specific test for metabolic intermediates produced when sugars are fermented to produce butylene glycol in addition to acids. Butylene glycol fermentation is one of the fermentation patterns exhibited by gram-negative rods. A positive VP test is almost never observed with a positive MR test because the MR test is based on the production of sufficient acid to lower the pH to below 4.5. If some of the sugar being fermented is used to make butylene glycol, then it cannot be used to produce acids. In this test you will add Barritt's reagents A and B to the medium and observe a color change after allowing sufficient time (about 10 minutes) for the reaction to occur.

CITRATE UTILIZATION

Simmons citrate agar is used to determine if a given isolate is able to utilize citrate. Citrate is the only oxidizable carbohydrate present. It also has a pH indicator that changes color if the citrate is used. The indicator (Brom thymol blue) turns from green to blue as the citrate is utilized. The reasons for the pH

changes are complicated and are related to the reaction of the alkaline earth metals in an aqueous medium. If these metals are used, alkaline by-products are produced which will result in a change in the medium from green to blue.

LABORATORY OBJECTIVES

- Understand the difference between fermentation and respiration and why a pH change is indicative of fermentation.
- Obtain an appreciation for the metabolic capabilities of bacteria and how these capabilities can be used to identify bacteria.
- Be introduced to the characterization of bacteria on the basis of their phenotypic characteristics.

MATERIALS NEEDED FOR THIS LAB

This exercise may be performed by groups or students as directed by your laboratory instructor. The following 24-hour cultures will be used for all sections of this exercise

> *Staphylococcus epidermidis*
> *Escherichia coli*
> *Enterobacter aerogenes*
> *Proteus vulgaris*
> *Bacillus subtilis*

FERMENTATION OF SUGARS

1. One tube of each media per organism:

> Phenol red glucose broth (Durham tube)
> Phenol red sucrose broth
> Phenol red lactose broth
> Phenol red mannitol broth

HYDROLYSIS OF STARCH

1. One starch agar plate per organism
2. Iodine solution.

METHYL RED–VOGES–PROSKAUER

1. One tube of MR-VP broth per organism.
2. Sterile test tube.
3. Sterile pipette and pipetter.
4. Reagents
> Methyl red indicator
> Barritt's Reagent A
> Barritt's Reagent B.

CITRATE UTILIZATION

1. One Simmons citrate agar slant per organism.

LABORATORY PROCEDURE

See Table 14.2 for a summary of all tests.

FERMENTATION OF SUGARS

Obtain and label the following tubes of media for each organism being tested: phenol red glucose broth (Durham tube), phenol red lactose broth, phenol red sucrose broth, phenol red mannitol broth.

1. Inoculate one of each of the sugar fermentation tubes with each organism. Aseptically transfer a loopful of culture to the tube.
2. Incubate all tubes at 37°C for 24 to 48 hours. If they must be incubated for longer than 48 hours, at 48 hours place them in the refrigerator until the next class period.
3. Following incubation, observe each tube for growth, the production of acid, and the production of gas in the glucose broth. Record your results on the Laboratory Report Form.

HYDROLYSIS OF STARCH

1. Obtain and label one Starch agar plate for each organism being tested.
2. Using aseptic technique, inoculate each plate with a single streak in the center of the plate.
3. Incubate all plates at 37°C for 24 to 48 hours.
4. Following incubation, flood the plates with iodine solution. Allow a few minutes for the color to develop. A clear (not stained purple) area around the bacterial growth indicates that starch has been hydrolyzed. If the purple color extends up to the growth, no hydrolysis occurred. You will find that, if you leave the iodine-flooded plates out in light, the purple coloration will fade. Be sure to record the results shortly after the addition of the iodine.
5. Record your results on the Laboratory Report Form.

METHYL RED–VOGES–PROSKAUER

1. Obtain and label one tube of MR–VP broth for each organism.
2. Inoculate each tube, being sure to use aseptic technique.
3. Incubate the inoculated tubes at 37°C for 24 hours.
4. Following incubation, transfer 1.0 ml of the MR-VP broth to a sterile test tube.
   a. To the larger volume, add 0.5 ml (about 10 drops) of methyl red indicator. If the indicator remains red, the test is positive; if it turns yellow, the test is negative.
   b. To the tube containing 1.0 ml of MR-VP broth, add 0.6 ml (about 12 drops) of Barritt's reagent A and 0.2 ml (about 4 drops) of Barritt's reagent B. Mix well and wait 10 minutes for the color to develop. A red color indicates a positive test; a yellow color indicates a negative test.
5. Record the results on the Laboratory Report Form.

CITRATE UTILIZATION

1. Obtain and label one tube of Simmons citrate agar for each organism being tested.
2. Using aseptic technique, inoculate the surface of the slant with your inoculating loop.
3. Incubate all tubes at 37°C for 24 to 48 hours.
4. Observe the Simmons citrate slants for color change. The presence of a blue color on the surface of the slant is a positive test.
5. Record your results on the Laboratory Report Form.

Table 14.2   Select Media for Determination of Carbohydrate Metabolism.

Media	Substrate(s)	Reagent(s)	Reaction and Description
Phenyl Red Broth	Glucose Lactose Sucrose Mannitol	Phenol red indicator	Acid (yellow) Gas (gas bubble in Durham tube) Alkaline (dark red color) No change (red)
Starch Agar	Starch	Iodine	Starch hydrolyzed (clear zone surrounding colony) Starch not hydrolyzed (no clear zone surrounding colony)
Methyl Red–Voges–Proskauer Broth	Glucose	Methyl red indicator	Acid production (red color) No acid production (non-red)
	Glucose	Alpha-naphthol KOH	Acetoin produced (red color) No acetoin produced (non-red)
Simmons Citrate Agar	Citrate	Brom thymol blue	Citrate utilized (growth, slant blue) Citrate not utilized (no growth, slant green)

NOTES_____

LABORATORY REPORT FORM

EXERCISE 14
PHYSIOLOGICAL CHARACTERISTICS OF BACTERIA:
CARBOHYDRATE METABOLISM

What is the purpose of this exercise?

1.   Record your results for the following tests:

ORGANISM	GLUCOSE			LACTOSE		SUCROSE		MANNITOL	
	Growth	Acid	Gas	Growth	Acid	Growth	Acid	Growth	Acid
Staphylococcus epidermidis									
Escherichia coli									
Enterobacter aerogenes									
Proteus vulgaris									
Bacillus subtilis									

ORGANISM	STARCH		METHYL RED		VOGES–PROSKAUER		CITRATE	
	Color	Reaction (+ or −)	Color	Reaction (+ or −)	Color	Reaction (+ or −)	Color	Reaction (+ or −)
Staphylococcus epidermidis								
Escherichia coli								
Enterobacter aerogenes								
Proteus vulgaris								
Bacillus subtilis								

QUESTIONS

1.   List two reasons for studying biochemical characteristics of bacteria.

      a.

      b.

2.   A student observes a tube of phenol red sucrose broth after a 24–hour incubation and notes that the tube is yellow. The tube is left in the incubator. When observed 3 days later, it appears dark red. What happened?

3.   How is it possible for an organism to grow on starch agar if it does not hydrolyze starch?

4.   Why is it unlikely that an organism will be positive for both the methyl red and the Voges-Proskauer tests?

PHYSIOLOGICAL CHARACTERISTICS OF BACTERIA: REACTIONS OF NITROGEN METABOLISM

In Exercise 14, we started to look at the various ways that bacterial organisms could utilize available nutrients for growth. In this exercise, we will look at how organisms can utilize nitrogen-containing compounds.

BACKGROUND

Bacteria normally metabolize carbohydrates as a source of energy, preferring to utilize proteins and amino acids for synthesis and growth. When, however, there is no oxidizable carbohydrate in the medium, or if the bacteria are unable to use the source that is present, the organism must utilize any available amino acids and proteins. Carbon chains are oxidized, and the unused, residual parts of the molecules, such as ammonia and hydrogen sulfide, are excreted. The following media restrict the availability of carbohydrates for bacterial growth. This forces the organisms to utilize, if possible, the available proteins and amino acids. We can then test for the by-products of protein decomposition or look for other evidence of proteolytic activity (the hydrolyzing or breakdown of protein).

HYDROLYSIS OF GELATIN

Gelatin is a protein colloid, maintaining its gel state at temperatures below approximately $25^{\circ}C$. The gel colloidal state is also dependent on the structural integrity of the proteins found in gelatin, and any hydrolysis of the peptides will result in destruction of the colloid. Organisms that produce the enzyme gelatinase can hydrolyze gelatin, with the degree of gelatin hydrolysis serving as an indicator of the activity of the enzyme. The resulting products are soluble peptides and amino acids that can serve as a source of carbon and energy for the organism. To test for gelatin liquefaction, it is only necessary to inoculate a tube of nutrient gelatin, incubate for 24 hours, and then determine if the gelatin will solidify when cooled. Failure to solidify (when an uninoculated control has solidified) is a positive test for gelatin hydrolysis.

An alternative method involves incubating the gelatin stab at room temperature for 5 to 7 days.

This method illustrates not only the presence of liquefaction, but also the pattern of liquefaction.

INDOLE PRODUCTION

Indole is a compound produced when the amino acid tryptophan is hydrolyzed to pyruvic acid (which is used for energy metabolism) and indole (which is excreted) through the action of tryptophanase. If indole is present, it reacts with Kovac's reagent to produce a red color. The reagent does not mix with the aqueous medium, forming instead a layer on the surface. The red color develops in that layer.

The presence of tryptophanase is a significant distinction between *Escherichia coli*, which produces indole and is therefore indole positive, and the other enteric organisms, which do not produce indole.

HYDROGEN SULFIDE PRODUCTION

Hydrogen sulfide (H_2S) is a by-product of the breakdown of cysteine by bacteria that produce the enzyme cysteine desulfurase. This enzyme enables the organisms to use the sulfur-containing amino acids for energy metabolism. Hydrogen sulfide is excreted and can be tested for by taking advantage of the formation of a black precipitate when it reacts with either silver, iron, or lead in the medium. To test for hydrogen sulfide production, you use a medium containing the sulfur-containing amino acids and a source of one of the metals. Iron (as ferrous sulfate) and lead (as lead acetate) are the most commonly used sources. The medium is inoculated as a stab into an agar deep. When the hydrogen sulfide is produced, a black precipitate forms along the line of inoculation.

UREA HYDROLYSIS

Urea is produced as a by-product of protein and nucleic acid decomposition. If an organism produces the enzyme urease, it enables bacteria to detoxify urea (a waste product) and to release usable energy as it hydrolyzes it to ammonia and carbon dioxide.

The ammonia reacts with the water in the medium to produce ammonium hydroxide, causing the pH to rise. Urea medium contains phenol red indicator in addition to urea, and has an initial pH of between 6.5 and 6.8. Refer back to Exercise 14 to see how the medium will change color as the ammonia is produced.

NITRATE REDUCTION

Nitrate reduction is characteristic of several organisms, resulting in the formation of nitrite. This is frequently used in diagnostic protocol for gram-negative bacilli. Nitrate is utilized as the final electron acceptor in anaerobic respiration, reducing the nitrate to nitrite. Chemicals are used to detect the presence of nitrite.

After nitrite is produced, some organisms will further reduce it to gaseous products (nitrogen gas) or to ammonia. The nitrogen gas is emitted to the atmosphere, and the ammonia is frequently used for the synthesis of proteins, although it may be excreted into the medium. If an organism tests negative for the presence of nitrite, it can mean one of three things:

1. the organism does not reduce nitrate;
2. the organism reduced nitrate to nitrite, but also produces nitrite reductase, which reduces the nitrite to ammonia; or
3. denitrification of the nitrate has occurred, resulting in the reduction of nitrite to nitrogen gas.

If the nitrate test is negative for the presence of nitrite, an additional test for the presence of unreacted nitrate should be performed. Zinc dust is added to the tube. This will catalyze the reduction of nitrate to nitrite. As the test reagents are still present in the tube, a red color will develop as nitrate is reduced to nitrite.

LABORATORY OBJECTIVES

*   Understand why some parts of amino acid molecules are excreted when the amino acids are used for energy metabolism.
*   Obtain an appreciation for the metabolic capabilities of bacteria and how this can be used to aid in the identification of bacteria.
*   Be introduced to the characterization of bacteria on the basis of their phenotypic characteristics.

MATERIALS NEEDED FOR THIS LAB

HYDROLYSIS OF GELATIN
1. Twenty-four–hour cultures of
    Enterobacter aerogenes
    Escherichia coli
    Proteus vulgaris
2. Nutrient gelatin (1 tube per organism)

INDOLE PRODUCTION
1. Twenty-four–hour cultures of
    Enterobacter aerogenes
    Escherichia coli
    Proteus vulgaris
2. Tryptone broth (1 tube per organism)
3. Kovac's reagent

HYDROGEN SULFIDE PRODUCTION
1. Twenty-four–hour cultures of
    Enterobacter aerogenes
    Escherichia coli
    Proteus vulgaris
2. Peptone iron agar (1 tube per organism)

UREA HYDROLYSIS
1. Twenty-four–hour cultures of
    Enterobacter aerogenes
    Escherichia coli
    Proteus vulgaris
2. Urea broth (1 tube per organism)

NITRATE REDUCTION
1. Twenty-four-hour cultures of
    Acinetobacter calcoaceticus
    Escherichia coli
    Pseudomonas aeruginosa
2. Nitrate broth (5 ml per tube) —1 per organism
3. Reagents:
    a. Nitrate A
    b. Nitrate B
    c. Zinc dust
4. Toothpicks

LABORATORY PROCEDURE
See Table 15.1 for a summary of all tests.

HYDROLYSIS OF GELATIN
1. Obtain and label one tube of nutrient gelatin for each organism being tested.
2. Inoculate the nutrient gelatin tubes by stabbing with the inoculating needle into the center of the tube.
3. Incubate at 37°C for 24 to 48 hours.
4. Following incubation, place the nutrient gelatin cultures, along with an uninoculated control, into a cold water or ice bath. After the control has solidified, tilt each tube to determine if the gelatin has remained liquefied. Any cultures that fail to solidify should be recorded as gelatinase positive.
5. An alternative method involves the incubation of cultures at room temperature for 5 to 7 days. This enables the determination of not only the presence of liquefaction but also its pattern.

6. Record your results on the Laboratory Report Form.

INDOLE PRODUCTION
1. Obtain and label one tube of tryptone broth for each organism being tested.
2. Inoculate the tubes with an inoculating loop or needle.
3. Incubate at 37°C for 24 to 48 hours.
4. Following incubation, add approximately 10 drops of Kovac's reagent to the tryptone broth. The reagent will form a layer on the surface and develop a red color if indole is present. The development of a red color should be recorded as positive for the production of indole.
5. Record your results on the Laboratory Report Form.

HYDROGEN SULFIDE PRODUCTION
1. Obtain and label one tube of peptone iron agar for each organism being tested.
2. Inoculate the tubes by stabbing with the inoculating needle into the center of the agar.
3. Incubate at 37°C for 24 to 48 hours.
4. Examine the peptone iron agar tubes for the presence of lead or iron sulfate. Any blackening of the medium indicates that hydrogen sulfide was produced.
5. Record your results on the Laboratory Report Form.

UREA HYDROLYSIS
1. Obtain and label one tube of urea broth for each organism being tested.
2. Inoculate the tubes with the inoculating loop or needle.

3. Incubate at 37°C for 24 to 48 hours.
4. Examine the tubes of urea broth. A hot pink to red color indicates that the urea has been hydrolyzed and ammonia excreted. This should be recorded as a positive test.
5. Record your results on the Laboraotry Report Form.

NITRATE REDUCTION
1. Obtain and label one tube of nitrate broth for each organism being tested.
2. Inoculate the tubes with an inoculating loop or needle.
3. Incubate at 37°C for 24 to 48 hours.
4. Add 3 drops of Nitrate Reagent A, then 3 drops of Nitrate Reagent B to each tube of nitrate broth. If nitrite is present, a distinctive red color will develop. This means that the organism is positive for nitrate reduction.
5. If a red color did not develop, add a small amount of zinc dust to the broth. If a red color develops following the addition of zinc, it indicates that nitrate, not nitrite, was present in the tube. The red color is due to the reduction of nitrate to nitrite by the zinc (a reducing agent). The test should be recorded as negative for nitrate reduction (the organism did not reduce the nitrate).
6. If no color develops following the addition of the zinc dust, then neither nitrate nor nitrite is present in the tube. The organism reduced the nitrate to nitrite and then to ammonia or nitrogen gas. The test should be recorded as positive for both nitrate and nitrite reduction.
7. Record your results on the Laboratory Report Form.

Table 15.1   Select Media for Determination of Nitrogen Metabolism

TEST	Substrate	Regent	Reaction and Description
Gelatin Hydrolysis	Gelatin	None	Gelatin hydrolysis (gelatin liquefies) Gelatin not hydrolyzed (gelatin solid)
Indole Production	Tryptophan	Kovac's reagent	Indole produced (reagent turns dark red) Indole not produced (no change)
Hydrogen Sulfide Production	Cysteine	Ferrous sulfate	H$_2$S produced (black precipitate) H$_2$S not produced (no black precipitate
Urea Hydrolysis	Urea	Phenol red	Urease positive (bright pink media) Urease negative (salmon media)
Nitrate Reduction	Nitrate	Sulfanillic acid Dimethyl-alphanaphthylamine	Nitrate reduced (red color) Nitrate not reduced (no red color) Nitrite reduced (no red color)
		Zinc dust	Nitrite reduced (gray color) Nitrate not reduced (red color)

NOTES_____

LABORATORY REPORT FORM

EXERCISE 15

PHYSIOLOGICAL CHARACTERISTICS OF BACTERIA: REACTIONS OF NITROGEN METABOLISM

What is the purpose of this exercise?

Record the results you obtained for the following tests and organisms:

ORGANISM	HYDROLYSIS OF GELATIN	INDOLE PRODUCTION	HYDROGEN SULFIDE PRODUCTION	UREA HYDROLYSIS	NITRATE REDUCTION
Acinetobacter calcoaceticus	■	■	■	■	
Enterobacter aerogenes					■
Escherichia coli					
Proteus vulgaris					■
Pseudomonas aeruginosa	■	■	■	■	

QUESTIONS

1. What is hydrolysis? List three examples.

2. Your microorganism grows well in the peptone iron agar, but does not form a black precipitate. What does this mean?

3.   Why is gelatin less suitable as a solidifying agent for bacteriological media than is agar?

4.   When changing a baby's wet diaper, why does it often smell of ammonia?

5.   In the nitrate reduction test, what does the presence of gas indicate?

PHYSIOLOGICAL CHARACTERISTICS OF BACTERIA: MISCELLANEOUS REACTIONS

BACKGROUND

Many bacteria produce enzymes that are secreted into the medium. For example, the hydrolysis of starch (Exercise 14) and the liquefaction of gelatin (Exercise 15) occur because some bacteria secrete hydrolytic enzymes that can depolymerize (break down into smaller units) these molecules. Many of the secreted enzymes have ecological or medical significance, and a few have diagnostic significance as well.

The clinical significance of some of these enzymes lies in their ability to attack substrates in host cell membranes, tissues, or body fluids, thereby interfering with host defense mechanisms or homeostasis. In other instances the enzymes are antigenic and can be tested for by serological reactions that provide evidence of exposure to the bacteria that produced the enzymes.

The enzymatic reactions we will examine in this exercise are ones that have been shown to be useful in the identification of unknown isolates. They are, of course, also important for the normal metabolism of the cell.

When bacteria produce flagella they are able to swim through liquids and soft media. They are said to be *motile*. Motility, as we will test for it here, is the result of the activity of flagella by those species of bacteria that produce them.

It is usually possible to test for enzymatic reactions by adding the enzyme's substrate to the bacterial culture and then testing for the appearance of the product or the disappearance of the substrate. For example, when you tested for starch hydrolysis in Exercise 14, you grew the bacteria on starch agar and then added iodine to test for the hydrolysis of starch.

SPECIFIC ENZYMATIC TESTS:
CATALASE TEST

Catalase is an enzyme produced by many organisms, including man. Indeed, it is so commonly produced that those organisms that do not produce it are rare enough so that the lack of catalase is a significant diagnostic characteristic. The enzyme converts hydrogen peroxide into water and oxygen and does so vigorously enough to cause foaming. (What happens when you use hydrogen peroxide to cleanse a cut?)

Hydrogen peroxide is produced when some organic molecules are directly oxidized. It must be rapidly removed because it is highly toxic to most cells—hence, most cells that are aerobic or facultative produce catalase, and those that do not are either obligate anaerobes or microaerophilic. Catalase is tested for by dropping one drop of hydrogen peroxide (the substrate) on a bacterial colony. If bubbles appear (the product), the organism is catalase positive. The catalase test is a very important one for differentiating between the genera *Staphylococcus* and *Streptococcus* and between the genera *Bacillus* and *Clostridium*.

OXIDASE TEST

Oxidase is an enzyme involved in certain oxidation reactions in aerobic bacteria. A relatively few genera, notably *Neisseria, Pseudomonas,* and a few other gram-negative rods, produce it, making it an important diagnostic test. To perform the test, it is necessary to combine the oxidase reagent (the substrate) with the colony being tested. If the colony is positive, it will turn first pink, then purple, and eventually black (the colored products are the result of the enzymatic activity). It is sometimes more convenient to simply flood the plate with the reagent and observe the color changes.

COAGULASE TEST

Coagulase is an enzyme produced by *Staphylococcus aureus*, which converts fibrinogen to fibrin, resulting in the coagulation of blood plasma. When coagulase is produced by a pathogenic organism, it results in the formation of a fibrin meshwork that serves to protect the organism from the host's natural defense mechanisms. The enzyme is not produced by *Staphylococcus epidermidis* or by any other gram-positive coccus. This test is very important because it allows for rapid identification of an important and commonly encountered pathogenic organism. A rapid screening test for coagulase may be completed by mixing a small amount of culture material with a drop of human or rabbit plasma. Try to emulsify the suspension. If the cells appear to agglutinate (clotting the plasma), the test is positive. Cells that are coagulase negative will produce an even, nonagglutinated suspension. A more definitive and sensitive test requires incubation of the bacterium in plasma at 37°C. Coagulation within 24 hours is considered positive for the presence of a virulent strain of *Staphylococcus aureus*.

DNase TEST

DNase is an excreted enzyme (like the gelatin and starch enzymes) that hydrolyzes DNA. A special medium, containing DNA and an indicator, is used to measure this activity. The bacterial species to be tested is spotted on the surface of the plate and the culture incubated for 24 hours. If DNA is hydrolyzed (by DNase), the medium will change color. DNase production is an important diagnostic characteristic for the staphylococci and some gram-negative rods.

ESCULIN HYDROLYSIS

Esculin hydrolysis (and growth in a medium containing bile) is considered to be definitive for the enterococci, a group of streptococci that are frequently encountered in clinical samples. This test uses a medium containing both esculin and bile. Positive organisms (enterococci) will grow and turn the medium black. The black color is due to a hydrolysis product of esculin that reacts with the iron in the medium. Bile is added as an additional test for the enterococci. It has no effect on the hydrolysis of esculin, except that it limits the type of organism that can grow on the medium—the organism must be resistant to bile.

MOTILITY

Whereas motility is not due to a chemical reaction, it is a characteristic of some organisms and can be readily observed by the use of appropriate media. Bacteria that have flagella are able to swim through liquids and soft media (less than half the amount of agar usually found in plated or slanted media). This ability can be observed directly with a wet mount slide, or indirectly by using soft agar deeps. (Wet mounts are discussed in Exercise 3.) Soft agar deeps are inoculated with a single stab down the center of the deep. Motile bacteria will move outward from the stab, resulting in a medium with a very hazy and cloudy appearance. Nonmotile bacteria cannot move and therefore grow only along the stab, producing a clear and sharp stab line. Some media contain a dye that is turned red by bacterial action, making the above observations much more distinct. When you examine your results, the difference will be clear and obvious.

LABORATORY OBJECTIVES

The enzymes secreted by bacteria have both ecological and clinical significance. The production of these enzymes may also be of diagnostic significance. You should

- Try to understand the ecological significance of excreted enzymes. For example, how do bacteria and other microbes cause the decomposition of dead plant and animal remains?
- Understand the possible clinical significance of certain bacterial enzymes (hemolysin, coagulase, collagenase, etc.). Those enzymes that are clinically significant may attack certain cells or inhibit certain processes in the host, or they may be serologically important.
- Appreciate the role that certain enzymes play in the metabolism and in the identification of bacteria.

MATERIALS NEEDED FOR THIS LAB

FOR CATALASE, OXIDASE AND COAGULASE TESTS
1.   Twenty-four–hour cultures of
         Lactococcus lactis
         Micrococcus luteus
         Pseudomonas aeruginosa
         Staphylococcus aureus
         Staphylococcus epidermidis.
2.   One TSA plate per organism.

CATALASE TEST
1.   Hydrogen peroxide
2.   Microscope slide
3.   Pasteur pipette
4.   Toothpick.

OXIDASE TEST
1.   Oxidase reagent or oxidase disk
2.   Filter paper
3.   Petri plate
4.   Toothpick.

COAGULASE TEST
1.   Coagulase plasma
2.   Sterile test tube
3.   Pipettes.

DNase TEST
1.   Twenty-four–hour cultures of
         Serratia marcescens
         Enterobacter aerogenes
2.   DNase Agar containing indicator, 1 plate for every 2 organisms to be tested.

ESCULIN HYDROLYSIS
1. Twenty-four–hour cultures of
     Enterococcus faecalis
     Streptococcus mitis
2. Bile esculin agar talls, 1 per organism

MOTILITY
1. Twenty-four–hour cultures of
     Enterobacter aerogenes
     Staphylococcus epidermidis
2. Motility agar tall, 1 per organism

LABORATORY PROCEDURES

FOR CATALASE, OXIDASE, AND COAGULASE
TESTS
1. Prepare a streak plate of each organism (see
   Exercise 9).
2. Incubate at 37°C for 24 to 48 hours.
3. Following incubation, perform the following
   tests on the organisms:

CATALASE TEST
1. Spot-catalase test: Place a drop of hydrogen
   peroxide on a clean microscope slide. Using a
   toothpick, carefully remove an isolated colony
   from the streak plate and add it to the drop of
   hydrogen peroxide.   Observe for bubbles.
   NOTE: Be sure to dispose of the toothpick as
   indicated by your instructor.
   OR
   Plate-catalase test: Carefully add a drop of
   hydrogen peroxide to an isolated colony on the
   streak plate. Observe for bubbles.
2. If bubbles are present, the organism is catalase
   positive.
3. Record your results on the Laboratory Report
   Form.

OXIDASE TEST
1. Spot-oxidase test: Place a piece of filter paper
   in an empty petri dish. Place a few drops of
   oxidase reagent on the filter paper. Using a
   toothpick, transfer an isolated colony from the
   streak plate to the moistened filter paper.
   Observe for a color change.
   OR
   Plate-oxidase test: Place one drop of the
   reagent on the colony to be tested. Observe for
   a color change.
2. A color change to pink, then purple indicates
   that the organism is oxidase positive.
3. Record your results on the Laboratory Report
   Form.

COAGULASE TEST
1. Slide screening test: Place two drops of
   coagulase plasma reagent in the center of a
   slide. Using a toothpick, transfer an isolated
   colony from the streak plate and emulsify it in
   the drop of plasma.   If the organism is
   coagulase positive, it will agglutinate, whereas a
   coagulase negative organism will emulsify
   easily and produce a uniformly hazy
   suspension.   NOTE: Your instructor may
   demonstrate this procedure. If you perform the
   test, be sure to dispose of the toothpick as
   indicated by your instructor.
   OR
   Tube incubation method:
     a. Using a sterile pipette, transfer 0.1 ml
         of coagulase plasma into a test tube.
     b. Add 0.4 ml of distilled water.
     c. Using your inoculating loop, transfer
         several colonies from your streak plate
         to the tube.
     d. Incubate at least 4 hours. A coagulase-
         positive organism will coagulate the
         entire volume of plasma within 24
         hours.
2. Record your results on the Laboratory Report
   Form.

DNase TEST
1. Using your marker, divide the DNase plate(s) in
   half. Draw a circle the size of a dime in the
   center of each half. Inoculate one organism on
   each half of the plate, spreading it out to fill the
   circle.
2. Incubate at 37°C for 24 to 48 hours.
3. If the organism has produced DNase, there will
   be decolorization of the dye surrounding the
   area of growth. This would be a positive test
   for DNase production.
4. Record your results on the Laboratory Report
   Form.

ESCULIN TEST
1. Using your inoculating needle, inoculate each
   tube of bile esculin tall with a single stab into
   the center of the tube.
2. Incubate the tubes at 37°C for 24 to 48 hours.
3. Observe the tubes. The presence of a black
   color along the line of inoculation indicates that
   esculin has been hydrolyzed. This reaction
   would be positive for the hydrolysis of esculin.
   If there is no black coloration produced, note
   whether the organism was able to grow in the
   presence of bile.
4. Record your results on the Laboratory Report
   Form. Use (+) to indicate a positive test; (−) to

indicate a negative esculin test; and (0) to indicate the absence of any growth.

MOTILITY

1. Inoculate the motility agar with a single stab into the center of the tube.
2. Incubate the tubes at 37°C for 24 to 48 hours.
3. Observe the growth along the stab line in the motility agar. If the organism is nonmotile, the stab will be clearly defined; if the organism is motile, the stab will be cloudy and hazy, indicating that the organism has moved through the medium.
4. Record your results on the Laboratory Report Form.

Table 16.1   Select Media for Determination of Miscellaneous Metabolic Activities

TEST	Substrate or Enzyme Detected	Reagents	Reaction and Description
Catalase Test	Enzyme catalase	Hydrogen peroxide	Catalase present (visible bubbles) Catalase not present (no visible bubbles)
Oxidase Test	Enzyme cytochrome c	Tetramethyl-phenylnediamine	Oxidase produced (deep violet or purple color) Oxidase not produced (no color change)
Coagulase Test	Enzyme coagulase	Defibrinated serum	Coagulase positive (clot formation) Coagulase negative (no clot formed)
DNase Test	DNase	DNA pH Indicator	DNase positive (color change in media surrounding bacterial growth) DNase negative (no color change present)
Esculin Test	Esculin	Ferric citrate	Utilize esculin in presence of bile (dark-brown color) Do not utilize esculin (no change, growth)
Motility Test	Presence of motility	Soft agar	Motile (movement of growth away from line of inoculation) Nonmotile (growth only occurs along line of inoculation)

Name _____

Section _____

LABORATORY REPORT FORM

EXERCISE 16
PHYSIOLOGICAL CHARACTERISTICS OF BACTERIA: MISCELLANEOUS REACTIONS

What is the purpose of this exercise?

Complete the following chart with your results:

ORGANISM	CATALASE TEST	OXIDASE TEST	COAGULASE TEST	DNase TEST	ESCULIN HYDROLYSIS	MOTILITY
Enterobacter aerogenes						
Enterococcus faecalis						
Lactococcus lactis						
Micrococcus luteus						
Pseudomonas aeruginosa						
Serratia marcescens						
Staphylococcus aureus						
Staphylococcus epidermidis						
Streptococcus mitis						

QUESTIONS

1. What is coagulase? How is it related to pathogenicity?

2.   What is the medical significance of DNase? of coagulase? of catalase?

3.   Why do obligate anaerobes not require catalase?

4.   A catalase test is performed on an isolated colony on a blood agar plate and bubbles appear. The test is repeated by performing a spot-catalase test on another colony from the same pure culture. This time there are no bubbles. Explain the difference in results.

MULTIPLE TEST MEDIA

We have seen how various tests/media can assist us in learning about the enzymatic processes undertaken by microorganisms. Each media would detect the presence of a single enzymatic process. This is extremely beneficial for learning about microbial metabolism, but it is extremely expensive and time-consuming if all of these individual tests have to be performed for every organism being tested. Multiple test media provide a more rapid and less expensive alternative.

BACKGROUND

Three of the more commonly used multiple test media are Litmus Milk, Triple Sugar Iron Agar (TSIA), and Sulfur, Indole, Motility (SIM) agar.

LITMUS MILK

Milk is an excellent growth medium for organisms and provides many varied test results. The skim milk in this medium provides organisms with a source of carbohydrate (lactose) and protein (casein). Litmus, in the form azolitmin, is present as a pH indicator. Litmus appears purple-blue when neutral, pink with the production of acid (pH 4.5), and blue with a pH over 8.

Lactose fermentation. The fermentation of lactose, which results in the formation of lactic acid, will result in the lowering of the pH of the medium. This causes the media to turn pink.

Alkaline reaction. Some organisms that are unable to utilize the lactose in the medium will use proteins in the milk. This results in the production of ammonia, which causes the tube to become very alkaline. The medium appears blue.

Clot formation. Two forms of clots can form in litmus milk.
a. Acid curd. Those organisms that utilize lactose produce lactic acid, which causes a lowering of the pH in the medium. This can result in the precipitation of casein and the formation of a hard or acid curd.
b. Rennin curd. Organisms that produce rennin can cause the production of a calcium-casein complex. This curd is a soft curd that retracts from the tube. Also often present is a grayish fluid or whey. (Remember Little Miss Muffet and her curds and whey?)

Gas production. Many of the enterics produce gas during the metabolism of lactose. Recall that an organism only produces gas from a carbohydrate when it has been fermented (tests positive for the production of acid). The gas is visible as breaks in the clot.

Litmus reduction. If the litmus serves as the electron acceptor during fermentation, it results in reduction of litmus, which is seen as the conversion of the medium to white in all but the top centimeter of the tube.

Peptonization. Proteolytic enzymes can digest the casein in the milk. This results in the formation of a clear liquid or watery appearance at the top of the tube.

TRIPLE SUGAR IRON AGAR (TSIA)

TSIA agar is used to differentiate among the gram-negative enteric organisms. It contains glucose, lactose, and sucrose; however, the glucose is only 1/10 the concentration of the lactose or sucrose. Phenol red and ferrous sulfate are present as indicators—the phenol red for acid production and the ferrous sulfate for hydrogen sulfide production.

Glucose fermentation. If glucose is the only carbohydrate being fermented, it results in the butt of the tube turning yellow owing to acid production.

Lactose or sucrose fermentation. If an organism ferments lactose and/or sucrose, the entire tube (butt and slant) will be yellow because of the increased production of acid. An organism that ferments lactose and/or sucrose will also have fermented glucose.

Gas production. If the organism produces gas from the glucose and/or lactose or sucrose, it will result in the formation of gas pockets or breaks in the agar.

Hydrogen sulfide production. Hydrogen sulfide can be produced along with the fermentation of glucose or glucose and lactose/sucrose. The presence of ferrous sulfide will cause a black precipitate throughout the media. If only glucose is fermented, the butt of the tube will be yellow, but the presence of the ferrous sulfide can mask the color of the butt. The slant in this case will still be red. If hydrogen sulfide is produced, one can automatically assume that the butt is yellow. If either one of the other sugars is fermented, the slant will be visibly yellow.

SULFUR, INDOLE, MOTILITY (SIM) AGAR
As implied by the name of this media, the production of hydrogen sulfide, indole, and motility can be tested using this media. Once again we see the presence of sodium sulfates, which can be converted to hydrogen sulfide, which then combines with the ferrous sulfate to form ferrous sulfide, giving us the expected black precipitate.

Indole production. The amino acid tryptophan is present in SIM agar as a source of pyruvic acid for organisms. If tryptophan is broken down to pyruvic acid, indole and ammonia are also formed. Kovac's reagent will combine with the indole present, producing the bright cherry-red layer at the media surface.

Motility. The motility of an organism can be seen by its growth pattern in SIM agar. This is a semi-solid agar that encourages the movement of motile organisms away from the line of inoculation. It is most readily observed when the organism also produces hydrogen sulfide, as the actual movement away from the site of inoculation can be seen. If the organism does not produce hydrogen sulfide, motility can be seen by comparing the cloudiness of the media to that of an uninoculated control.

LABORATORY OBJECTIVES
Multiple test media can be used to examine several enzyme systems with a single inoculation. By using these multi-test media you should be able to
- distinguish between the varied results present in a single medium;
- compare the results obtained by the single test system and the multiple test system.

MATERIALS NEEDED FOR THIS LAB

Twenty-four-hour to 48–hour cultures of the following:

   Escherichia coli
   Proteus vulgaris
   Pseudomonas aeruginosa
   Shigella flexneri
   Salmonella typhi
   Micrococcus luteus

A. LITMUS MILK
1. One tube of litmus milk for each organism being tested.

B. TRIPLE SUGAR IRON AGAR (TSIA)
1. One tube of TSIA for each organism being tested.

C. SULFUR, INDOLE, MOTILITY (SIM) AGAR
1. One tube of SIM agar for each organism being tested.
2. Kovac's reagent

LABORATORY PROCEDURES

A. LITMUS MILK
1. Inoculate each of your litmus milks.
2. Incubate at 37°C for 24 to 48 hours.
3. Observe the tubes for any changes (see Table 17.1). You may want to incubate the tubes for an additional 24 to 48 hours.
4. Record your results on the Laboratory Report Form, Part A.

B. TRIPLE SUGAR IRON AGAR (TSIA)
1. Inoculate each slant with an inoculating needle. Stab straight down to the butt or the tube. As the needle is removed from the butt, inoculate the slant.
2. Incubate at 37°C for 24 to 48 hours.
3. Observe the tubes for any changes (see Table 17.1).
4. Record your results on the Laboratory Report Form, Part B.

C. SULFUR, INDOLE, MOTILITY (SIM) AGAR
1. Inoculate each tall with an inoculating needle. Stab straight down to the butt of the tube.
2. Incubate at 37°C for 24 to 48 hours.
3. Observe the tubes for any changes (see Table 17.1).
4. Add Kovac's reagent to the top of the tall.
5. Record your results on the Laboratory Report Form, Part C.

Figure 17.1   Select Multiple Test Media.

TEST	SUBSTRATES	ENZYMES	REAGENTS	REACTION AND DESCRIPTION
Litmus Milk	Lactose	Lactase	Litmus pH indicator	Acid production (pink) Inability to use lactose (blue) Gas production (break in clot)
	Casein	Casease + lactase	Litmus pH indicator	Acid clot formation (solid)
		Proteolytic enzymes		Clearing of fluid
		Rennin	Litmus pH indicator	Soft curd, whey (custard like)
		Casease	Litmus pH indicator	Alkaline reaction resulting in ammonia production (darker blue)
	Litmus		Litmus pH indicator	Litmus reduction (white) Litmus not reduced (blue → pink)
Triple Sugar Iron Agar (TSIA)	Glucose	Phosphoglucomutase	Phenol Red pH indicator	Acid production (butt of tube turns yellow, slant remains red) No fermentation (entire slant red)
	Lactose	Lactase	Phenol Red pH indicator	Entire slant acid (+ glucose utilization)
	Sucrose	Sucrase	Phenol Red pH indicator	Entire slant acid (+ glucose utilization)
	Carbohydrate fermentation			Gas production (slant breaks up)
	Cysteine	Cysteine desulfurase	Iron salt	Blackening of medium (iron sulfide)
Sulfur, Indole, Motility Agar (SIM)	Cysteine	Cysteine desulfurase	Iron salt	Blackening of medium (iron sulfide)
	Tryptone	Tryptophanase	Kovac's reagent	Indole production (red layer)
	Motility			Iron sulfide diffuse through media Cloudiness of media away from stab

NOTES_____

Name _____

Section_____

LABORATORY REPORT FORM
EXERCISE 17
MULTIPLE TEST MEDIA

What is the purpose of this exercise?

A.   LITMUS MILK RESULTS

ORGANISM	ACID	ALKALINE	NO CHANGE	CURD	PROTEOLYSIS	LITMUS REDUCTION	GAS
Escherichia coli							
Proteus vulgaris							
Pseudomonas aeruginosa							
Shigella flexneri							
Salmonella typhi							
Micrococcus luteus							

B.   TRIPLE SUGAR IRON AGAR (TSIA)

ORGANISM	BUTT	SLANT	GAS PRODUCTION	H$_2$S PRODUCTION
Escherichia coli				
Proteus Vulgaris				
Pseudomonas aeruginosa				
Shigella flexneri				
Salmonella typhi				
Micrococcus luteus				

C.   SULFUR, INDOLE, MOTILITY (SIM) AGAR

ORGANISM	H$_2$S PRODUCTION	INDOLE	MOTILITY
Escherichia coli			
Proteus vulgaris			
Pseudomonas aeruginosa			
Shigella flexneri			
Salmonella typhi			
Micrococcus luteus			

QUESTIONS

1.   How did the results with the multiple test media compare to the results for the same organism in the single test media?

2.   What are the advantages of using a multiple test media? Single test media?

3.   Is it possible to have the following results? Why or why not?

     a.   With TSIA, slant yellow and butt red

     b.   With Litmus Milk, purple and clot

     c.   With SIM, motility and no hydrogen sulfide

DEVELOPMENT AND USE OF DIAGNOSTIC KEYS

In the previous exercises, you have been learning about the varied enzyme systems that given organisms possess through an examination of the metabolic processes they perform and the waste products formed. When this information is combined with basic morphological and staining results it becomes possible to begin the identification process. In this exercise you will become familiar with the use of *Bergey's Manual of Determinitive Bacteriology*.

BACKGROUND

Diagnostic keys are commonly used for the identification of organisms. The characteristics of an unknown microorganism are compared to those of a wide variety of other organisms in an attempt to determine the genus and species identification of the unknown organism. It must be noted that this is **not** the same as classification of organisms. In classification, organisms are placed in taxonomic groups based on similarity of structure, whereas identification places an organism into a known taxonomic group based on its characteristics.

Classification is a complex process, often utilizing highly sophisticated testing procedures, such as DNA hybridization. Identification is a much simpler process, one not as concerned with the total characterization of the organism as it is with the utilization of a few easily determined test results, which will enable the distinction between varied groups of microorganisms. Classification methods are most commonly employed in research laboratories working with possible new microorganisms. Identification methods are most commonly used in the clinical laboratory where rapid, accurate, and inexpensive methods must be used for microbe identification.

The major reference text for classification and identification has been *Bergey's Manual of Determinitive Bacteriology*. As more and more information has become available, *Bergey's Manual* has become more of a reference for classification and nomenclature and less useful as an identification tool. The 9th edition of the book was retitled *Bergey's Manual of Systematic Bacteriology* when it was revised in the 1980s. This four-volume series discusses the classification and nomenclature of organisms, giving detailed descriptions of species and providing methods for enrichment and isolation. In 1994, *Bergey's Manual of Determinative Bacteriology* was published. It is not an abridged version, but rather was established to deal specifically with the identification of organisms.

The classical approach to identification is the development of *keys* or *diagnostic tables*. Included are the use of dichotomous keys, profile analyses, and numerical identification. All of these methods utilize characteristics similar to those you have obtained in the previous exercises, including:

Cell morphology
Staining reactions
Colony morphology
Biochemical characteristics.

THE DICHOTOMOUS KEY

A dichotomous key is a list of increasingly specific phenotypic traits that will allow you to assign a genus and species name to your organism by a stepwise elimination of other organisms. In this system, the identifying characteristics are used one at a time to eliminate unrelated organisms.

It is, in effect, a road map that takes you to the genus and species name of your unknown organism. It requires that you be able to define your isolate by answering simple, single-variable questions about it. Usually a dichotomous key's questions can be answered with a simple yes or no. For example, "Does the unknown organism ferment lactose?" or "Does the unknown organism hydrolyze gelatin?" In other cases the question calls for a choice between two mutually exclusive conditions or states, such as "Is it gram-positive or gram-negative?" or "Is it a bacillus or a coccus?" When you put a dichotomous key into diagram form, it should look like a pyramid, the first test being at the top or apex, and the last tests and the names of the identified bacteria at the base.

In order to use a dichotomous key, you need only know which tests must be performed and then follow the key to determine the correct genus and species. Obviously, however, someone must have first determined which tests could be applied diagnostically and then determined the best sequence in which to use the tests. As you examine the sample key, note that the tests applied late in the key are much more specific than those used early in the key. The later tests usually distinguish single species, whereas the early ones typically distinguish genera or even groups of genera.

The disadvantages of dichotomous keys include the necessity to interpret the tests absolutely a

Table 18.1   Data Matrix Chart—An Example

NAME OF ORGANISM	GRAM REACTION	CELL SHAPE	FERMENTS LACTOSE	LIQUEFIES GELATIN	COAGULASE TEST
E. coli	−	Bacillus	+	−	−
S. aureus	+	Coccus	+	+	+
S. epidermidis	+	Coccus	−	−	−
Ps. aeruginosa	−	Bacillus	−	+	−

as either plus/minus (+/−) or yes/no. There is no accommodation of variant organisms, or those with atypical characteristics differing from the type strain. It is also true that a single error in test interpretation can result in misdiagnosis. Each dichotomous key is based on judgments about the proper sequence of tests used in the key. In a very real way, the dichotomous key is like a road map—there may be several routes to the same destination, but a wrong turn early in the route will leave you lost in a strange neighborhood.

AN EXAMPLE EXERCISE: DEVELOPMENT OF A DICHOTOMOUS KEY

A bacterial characteristic data matrix has been compiled using the same characteristics as those you have been determining in the previous exercises. That data will be used to write your own dichotomous key. In the example given here, all the necessary information is provided (see Table 18.1). You should try to understand that the three sample charts are simply three ways of presenting the same data.

The matrix you will be using is more extensive than the example shown here because you will be using data representative of all the tests performed in the previous exercises. You will use that data to develop your dichotomous key.

For the organism shown in the example, the first branch in the key might be the gram reaction and/or the shape of the cells. The bacilli can then be differentiated on the basis of the formation of lactose, and the cocci can be distinguished by either the coagulase reaction or the liquefaction of gelatin.

The information contained in the Data Matrix Chart is presented in Figure 18.1 as it might appear in a flow diagram or as a "road map." Try to determine the relationship between the two ways of presenting the data.

The first branch in your key should be a characteristic that divides the group of organisms into two large, approximately equal groups. At each subsequent branch of your key, you should use increasingly more specific tests until individual organisms are identified. When you have completed the key, you should be able to use it to identify any of the bacteria that you worked with in the previous exercises (see Figure 18.1).

There is yet another way of presenting the dichotomous key. It takes the form of a series of statements about the organisms and is probably the form that you will most likely encounter. Table 18.2 is a short dichotomous key using the examples in this exercise (see Table 18.2).

PHENOTYPIC PROFILE ANALYSIS

Profile analysis is a diagnostic procedure that simultaneously compares several tests to develop a profile of the unknown and compares it with profiles in a reference data bank. The advantages of profile analysis include the ability to compare many tests at once, avoiding the problem created by "wrong turns," and the ability to accommodate variants that have one or more atypical characteristics (see Table 18.3). The disadvantage of this technique is that, as additional characteristics are added to the profile, it is increasingly difficult to keep everything straight. Although the average human mind can reasonably handle up to 10 or so traits, as many as 20 or 30 individual traits must often be used to complete the identification. The development of computers has made profile analysis readily available. Several commercial systems are on the market. We will use one of these in Exercise 20.

PHENOTYPIC PROFILE ANALYSIS—A BRIEF DISCUSSION

If we use the example given above, a profile analysis system would consider all five of the given characteristics simultaneously. An organism that gave the results shown in Table 18.3 (the unknown organism) would be matched with *S. aureus*.

Figure 18.1   Dichotomous key—An example.

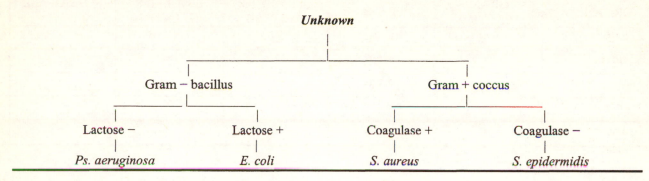

If one test varied from the reference profile, the system would report the goodness of fit (e.g., "4/5 match") with *S. aureus*.   In this case, however, because there are only two tests differentiating *S. aureus* from *S. epidermidis*, the profile analysis would alert you if the atypical test was one of the two needed to differentiate between these two species.

LABORATORY OBJECTIVES

To become familiar with the use of phenotypic characteristics as tools for bacterial identification, you will:

- Use phenotype characteristics in a data matrix chart to develop a diagnostic key that would allow anyone to identify any of the bacteria studied so far.
- Gain an understanding of how the identification of unknown bacterial isolates must follow a logical, step-by-step process.
- Gain an understanding of the differences, both advantageous and disadvantageous, between a dichotomous key and a phenotypic profile.
- Become familiar with the use of *Bergey's Manual of Determinative Bacteriology* for the identification of bacterial species.

LABORATORY PROCEDURES

1. Use the data matrix chart found in Table 18.4 to develop your own dichotomous key. Remember there is more than one way that this key can be put together—and not one of them is the "right" way!

2. You might simplify your task by starting with certain considerations:
    a. Can a separate key be used for major groups of bacteria, such as the gram-negative cocci, the gram-positive cocci, and so on?
    b. Determine which reactions divide the largest number of organisms into manageable groups.   For example, lactose fermentation, oxidase reaction, or hydrogen sulfide production might be used to separate the gram-negative bacilli into manageable groups, especially if the reactions are used sequentially.
    c. Use this approach to identify a "practice" unknown organism.
3. Try using the key on two or three of your known organisms to see if it works. Better yet, have another student try it!
4. Make sure that there is no ambiguity in the key. If a test is used more than once, consider if it could be moved up to a branch closer to the start.

Table 18.2   The Dichotomous Key

1. Organism is a gram-negative bacillus	If not, go to step 2.
a. Ferments lactose	*E. coli*
b. Does not ferment lactose	*Ps. aeruginosa*
2. Organism is a gram-positive coccus	
a. Organism is coagulase positive	*S. aureus*
b. Organism is coagulase negative	*S. epidermidis*

Table 18.3   Data Matrix Chart—An Example

ORGANISM	GRAM REACTION	CELL SHAPE	FERMENTS LACTOSE	LIQUEFIES GELATIN	COAGULASE TEST
E. coli	−	bacillus	+	−	−
S. aureus	+	coccus	+	+	+
S. epidermidis	+	coccus	−	−	−
Ps. aeruginosa	−	bacillus	−	+	−
Unknown	+	coccus	+	+	+

Table 18.4   Dot Matrix Chart for the Determination of the Identity of an Unknown Microorganism.

Organism	Shape	Gram Reaction	Endospores	Acid-Fast	Flagella	Oxygen	Glucose Fermentation	Sucrose Fermentation	Lactose Fermentation	Mannitol Fermentation	Methyl Red Test	Voges-Prauskauer	Citrate	Gelatin Liquefaction	Indole Production	Hydrogen Sulfide	Urease	Nitrate Reduction	Catalase	Oxidase	Coagulase	DNase	Esculin Hydrolysis
Acinetobacter calcoaceticus	B	–	–	–	–	A	+						+	–				+	+	–			
Bacillus cereus	B	+	+	–	+														+				
Bacillus megaterium	B	+	+	–	+	A	+											d	+	d			
Bacillus subtilis	B	+	+	–	+	A	+												+				
Citrobacter freundii	B	–	–	–	+	F	+G	d	d	+	+	–	+	–	–		d	+	+	–			–
Clostridium sporogenes	B	+	+	–	+	AN													–				
Corynebacterium xerosis	B	+	–	–	–	F	+	+	–		–			–	–				+	–		–	–
Enterobacter aerogenes	B	–	–	–	+	F	+G	+	+	+	–	+	+	w	–	–	–	+	+	–		–	+
Enterococcus faecalis	C	+	–	–	–	F	+	+	+	+									–				
Escherichia coli	B	–	–	–	+	F	+G	d	+	+	+	–	–	–	+	–	–	+	+	–		–	d
Klebsiella pneumoniae	B	–	–	–	–	F	+G	d	d	A	+		d		–	–	d	+	+	–		–	d
Lactococcus lactis	C	+	–	–	–	F	+		+	–									–	–			
Micrococcus luteus	C	+	–	–	–	A	–		–				–	+				–	+	+			–

Table 18.4, continued

	Shape	Gram Reaction	Endospores	Acid-Fast	Flagella	Oxygen	Glucose Fermentation	Sucrose Fermentation	Lactose Fermentation	Mannitol Fermentation	Methyl Red Test	Voges-Prauskauer	Citrate	Gelatin Liquefaction	Indole Production	Hydrogen Sulfide	Urease	Nitrate Reduction	Catalase	Oxidase	Coagulase	DNase	Esculin Hydrolysis
Micrococcus roseus	C	+	–	–	–	A	+		–				–	–				+	+	–			–
Moraxella catarrhalis	C	–	–	–	–	A	–	–	–	–				–	–		–	+	+	+			
Morganella morganii	B	–	–	–	+	F	+G	–	–	–	+	–	–	–	+	–	+	+	+	–		–	–
Mycobacterium phlei	B	W	–	+	–	A				+			+					+	+				
Mycobacterium smegmatis	B	W	–	+	–	A	+			+				+	+								
Proteus mirabilis	B	–	–	–	+	F	+G	–	–	–	+	d	d	+	–	+	+	+	+	–		d	–
Proteus vulgaris	B	–	–	–	+	F	+	+	–	–	+	–	–	+	+	+	+	+	+	–		d	d
Pseudomonas aeruginosa	B	–	–	–	+	A	+		–	+				+				+	+	+			
Salmonella typhimurium	B	–	–	–	+	F	+				+		+		–	+	–	+	+	–			
Serratia marcescens	B	–	–	–	+	F	+	+	–	+	–	+	+	+	–	–	–	+	+	–		+	
Shigella flexneri	B	–	–	–	–	F	+	–		+	+		–	–	d		–	+	+	–		–	–
Staphylococcus aureus	C	+	–	–	–	F	+	+	d	d		d		+			w	d	+	–	+	+	
Staphylococcus epidermidis	C	+	–	–	–	F	+	+	d	–		+		–			+	w	+	–	–	–	

Table 18.4, continued

	Shape	Gram Reaction	Endospores	Acid-Fast	Flagella	Oxygen	Glucose Fermentation	Sucrose Fermentation	Lactose Fermentation	Mannitol Fermentation	Methyl Red Test	Voges-Prauskauer	Citrate	Gelatin Liquefaction	Indole Production	Hydrogen Sulfide	Urease	Nitrate Reduction	Catalase	Oxidase	Coagulase	DNase	Esculin Hydrolysis
Streptococcus lactis	C	+	–	–	–	F			A	–									–				
Streptococcus mutans	C	+	–	–	–	F			+	+		+							–				+
Streptococcus pneumoniae	C	+	–	–	–	F			+	–		–							–				d
Vibrio parahaemolyticus	B	–	–	–	+	F	+	–	–	+	+	–	–	+	+	–	–	+		+		+	–

KEY: B = bacillus; C = coccus; A = aerobe; AN = anaerobe; F = facultative anaerobe; G = gas; w = weakly positive; d = variable

NOTES

LABORATORY REPORT FORM

EXERCISE 18
DEVELOPMENT AND USE OF DIAGNOSTIC KEYS

What is the purpose of this exercise?

Develop a diagnostic key for the organisms in Table 18.4. Remember that, as with any road map, there is more than one way to get to any destination. Attach your key to the Laboratory Report Form.

QUESTIONS

Use *Bergey's Manual of Determinative Bacteriology*, 9th Edition to answer the following questions.

1. Where would you find a description of the Archaeobacteria? _____

2. In which of the four major categories of bacteria are the facultatively anaerobic gram-negative rods found?

3. Where are prokaryotes described? _____

4. Find the pages that give a general approach to using the manual. _____

5. Where are the following **described**?
    a.   Enterobacteriaceae _____

    b.   Rickettsia _____

    c.   *Staphylococcus* _____

    d.   *Treponema* _____

6. What was the genus *Lactococcus* previously known as? What is its type species?

7. Where would you look to differentiate the endospore-forming bacteria? _____

USE OF DIAGNOSTIC KEYS:
BACTERIAL UNKNOWNS

One of the reasons you need to be familiar with the development of diagnostic keys is that you may have to apply that information to the solving of one or more unknowns. In clinical microbiology, every specimen sent to the laboratory for analysis is an unknown. It is often necessary to identify nonclinical isolates because pathogenic bacteria are frequently found in environmental samples and in contaminated food and water. In short, when you send in a culture from any source to the laboratory asking personnel to identify the predominant organism(s), you have presented them with an unknown.

BACKGROUND

It is not usually possible to identify bacteria simply on their morphology. Physiological characteristics provide a much more diverse set of criteria for identifying bacterial organisms. By combining both morphological and physiological characteristics, the identity of most organisms can be determined. S. T. Cowan and J. Liston in the 8th edition of *Bergey's Manual* propose the following steps in identifying an unknown:

1. Make sure that you have a pure culture.
2. Work from broad categories down to a smaller, specific category of organisms.
3. Use all the information available to you so as to narrow the range of possibilities.
4. Apply common sense at each step.
5. Use the minimum number of tests to make the identification.
6. Compare your isolate to type or reference strains of the pertinent taxon to ensure the identification scheme being used actually is valid for the conditions in your particular laboratory.

When difficulty is encountered in determining the identity of an unknown, you must check its purity, the appropriateness of the tests performed, the reliability of your techniques, and your correct use of the keys available. It should be noted that the most frequent mistakes in unknown identification is error in determining shape, gram reaction, and motility, or is due to contamination of the original culture.

PROBLEM-SOLVING APPROACH TO UNKNOWNS

The first procedures used in any identification protocol must be those that isolate the organisms into pure cultures. The technician must be familiar with the bacteria likely to be encountered and be able to recognize those bacteria by their colonial morphology. This initial isolation is sometimes referred to as primary isolation.

After isolation is achieved, the identification of the unknown proceeds by an orderly and stepwise elimination of alternate choices. The first choices are on the basis of the cellular morphology and staining reactions of the isolated culture. At the very least, a gram stain is considered essential, followed by whichever other stains would prove helpful (e.g., acid-fast and metachromatic granule stains).

Once the cell morphology, the staining reactions, and the colonial morphology are known, an educated guess often can be made as to the possible genus and species of the bacterium. Then, depending upon the possibilities suggested by the information already available, specific biochemical tests are used to confirm the identity of the unknown or to determine the genus and species if it is not already known. Which biochemical tests do you use? That is where your knowledge of and experience with diagnostic keys will prove useful. A good diagnostic protocol will use the least number of tests to arrive at the solution. To do this, you must know which tests are important for the type of organisms you are trying to identify. For example, you would use a different set of tests for gram-positive cocci than you would use for gram-negative bacilli.

The diagnostic key that you developed in the previous exercise may be used to identify your unknowns or to quickly determine if the unknown you are working with is not one of the species included in the key.

The preferred way to solve an unknown is to carefully determine which tests should be used, and in what order. The worst way, and often the most confusing way, to handle an unknown is to simply subject it to every available test, hoping that sufficient data will be generated to provide the correct answer. It is true that all of the needed data will have been generated, but because there was no careful stepwise elimination of alternative choices, you will usually be confronted with what often seems to be hopeless confusion.

PROCEDURE FOR SOLVING UNKNOWNS

First Step

Primary isolation: Streak plates for initial isolation and determination of colonial morphology.

Second Step

Staining reactions: Microscopic examinations for determination of cellular morphology and staining reactions

Third Step

Biochemical tests: Confirmation and identification by biochemical characterization, using a carefully determined set of tests

In clinical applications, the identification of the bacteria in submitted samples is usually expected within 72 hours, and often within 48 or less hours. Speed and accuracy are essential and can only be maintained if a planned protocol is followed. If you are working with a gram-negative bacillus, you do not use tests that are useful for the identification of gram-positive cocci. If a gram-positive bacillus that looks like a diphtheroid is observed to produce metachromatic granules, you do not need to do the oxidase test, but you do need to use a few selected tests to confirm the identification of *Coryne-bacterium diphtheria*. Conversely, if you isolated a gram-positive coccus, you would consider using the coagulase test, catalase test, mannitol fermentation, and gelatin liquefaction (in that order?).

In summary, the best approach to identifying a bacterial unknown is a problem-solving approach. After each step you must ask yourself "What possibilities are indicated by the results?" and "What tests are needed to further differentiate this isolate?" The surest way to become completely lost and confused is to attempt to apply tests in a shotgun fashion, hoping to hit upon the answer. Plan your approach carefully.

You may be required to write a complete laboratory report outlining the steps taken to determine the identification of your unknown organism(s). Guidelines for writing a laboratory report are outlined in Appendix A.

LABORATORY OBJECTIVES

You will be asked to identify one or more unknown bacterial organisms. To successfully accomplish that task, you must:

- Apply all of the skills learned in the previous exercises, including isolation of pure cultures, microscopic examination of stained smears, and recognition of colonial morphologies.
- Use a problem-solving approach to develop a protocol for the orderly application of diagnostic tests and the interpretation of the results obtained from those tests.
- Learn to anticipate what is next when the results of one set of tests are known.

MATERIALS NEEDED FOR THIS LAB

NOTE: The bacterial cultures include, but are not limited to, bacteria used in the previous exercises.

1.   Unknown culture: Your unknown specimen represents either a urine or a wound infection and contains two organisms.
2.   TSA plates.
3.   All routine staining reagents.
4.   Media and chemical reagents as needed to identify the unknown organisms.

LABORATORY PROCEDURE

HINTS AND SUGGESTIONS

1.   **Never** discard a culture until any subcultures of it have grown adequately for use.
2.   **Always** make a streak plate of your isolated cultures (even if they were derived from a single, well-isolated colony).
3.   **Always** keep a culture of your unknown isolates (a slant is preferred) until you are completely finished with them. You may need to refer back to them.
4.   **Always** keep all slides of your unknowns until you are completely finished with them.
5.   **Always** include controls in your tests.
6.   **Always** keep detailed notes of your laboratory work and in a secure place. It is almost always necessary to go back over them to determine some previous test results or to see if the results were interpreted correctly.
7.   **Always keep calm.**

IDENTIFICATION PROCEDURES

1.   The protocol you will follow for this laboratory exercise will be one you develop yourself. You must decide upon the protocol as you acquire data about your unknowns. You must decide what to do next.

2. The following initial steps are recommended:
   a. Make Gram stains and streak plates of your culture.
   b. On the basis of the Gram stains, and while you are waiting for the streak plates to grow, plan what tests will be used. Consult your keys from Exercise 18.
   c. Be able to tell your instructor why each desired media is necessary in the identification of your unknowns.
3. Continue with the testing until you have identified each organism in your unknown culture.

4. Your results may be added to the end of Table 18.4, filling in only those tests performed on the unknown.
5. Complete the Laboratory Report Form for each organism isolated, filling in only those tests performed in the identification process. Include under NOTES the sequence of tests performed as well as the reason(s) for choosing which test was done next.

NOTES _____

LABORATORY REPORT FORM

EXERCISE 19
USE OF DIAGNOSTIC KEYS: BACTERIAL UNKNOWNS

What is the purpose of this exercise?

UNKNOWN NUMBER _____   SOURCE _____

UNKNOWN ORGANISM(S) TEST RESULTS	ORGANISM #1	ORGANISM #2
Cellular Morphology and Staining Reactions:		
Cell shape		
Gram–stain reaction		
Endospore formation		
Acid-fast reaction		
Capsule formation		
Flagella		
Metachromatic granules		
Other morphological features		
Colonial Morphology:		
Colony pigmentation		
Color of agar around colony		
Shape of colony		
Colony margin		

UNKNOWN ORGANISM(S) TEST RESULTS	ORGANISM #1	ORGANISM #2
Carbohydrate Reactions:		
Glucose fermentation		
Lactose fermentation		
Sucrose fermentation		
Mannitol fermentation		
Hydrolysis of starch		
Methyl red test		
Voges-Proskauer test		
Utilization of citrate		
Thioglycollate growth (aerobic/anaerobic)		
Protein and Other Reactions:		
Gelatin hydrolysis		
Indole production		
Hydrogen peroxide production		
Catalase reaction		
Oxidase reaction		
Coagulase test		
Urea hydrolysis		
Nitrate reduction		
DNase activity		
Motility		
IDENTIFICATION OF ORGANISM		

NOTES: Include the sequence of tests performed (your protocol) as well as the basis for deciding which tests to perform.

UNKNOWN #1:

UNKNOWN #2:

QUESTIONS

1. What was the basis of your final determination for the identity of:

    Unknown #1?

    Unknown #2?

2. What would have been the effect if there were errors in any of the test results?

RAPID DIAGNOSTIC TESTS: PROFILE ANALYSIS

In Exercise 18 you developed a dichotomous key for the identification of bacterial isolates. You should recall that one of the problems with such keys is that they do not allow for consideration of variant strains. Another problem is the time involved. The clinical microbiology laboratory must be able to determine the identity of an unknown organism as quickly as possible.

Profile analysis identification, by estimating the most likely match rather than an exact match, does allow consideration of variant strains. In addition, several profile analysis systems have been developed that enable the rapid identification of unknowns.

BACKGROUND

A profile analysis uses the results of a selected series of tests to compare your isolate with a "library" of organisms whose biochemical characteristics are known. All of the tests receive approximately equal weighting and, therefore, no single test result can significantly alter the overall profile. Because profile analysis requires only a "best fit," rather than an exact fit, a 100% agreement between the unknown organism and the known organism is not required. Of course, if a 100% fit is possible, all the better. Most profile analyses report the degree of "fit" and reject those with less than some predetermined level (such as 90%).

Comparing the profile of the unknown with those of the organisms in the "library" can be a tedious job. Imagine searching through a long table of +'s and −'s until you found one that matched. Fortunately, such a tedious search is not necessary. Most profile analysis systems can be easily adapted to a computer analysis or to a relatively simple mathematical coding process that greatly simplifies the comparison process.

In developing rapid diagnostic test systems, accuracy and reliability must not be sacrificed. One factor that has helped maintain accuracy is the elimination of any culturing beyond the initial isolation of the organism. This eliminates much of the potential for contamination.

MULTIPLE, RAPID TEST AND AUTOMATED SYSTEMS

Two major approaches have appeared to aid in providing not only rapid but also accurate diagnosis of organisms. The first is the utilization of multiple test systems. These incorporate a number of media into a single unit, enabling the performance of multiple inoculations in a very short period of time. Incubation of 24 to 48 hours, however, is often still required.

Another approach is that of "rapid" tests — systems that provide reliable results in 2 to 4 hours. Because this is not long enough for significant growth to occur, these systems use varied substrates that can utilize preformed or constitutive enzymes always present in the organisms being tested.

Automated methods are the quickest and most accurate ways for bacterial identification. They are also the most expensive methods. Hospitals, where speed and accuracy are more important that cost, are the major users of automated systems.

The advantages of these systems include (1) their short incubation time; (2) direct inoculation from an isolated colony; (3) ease of handling; (4) cost-effectiveness; (5) minimal storage space; (6) ease of adjustment to taxonomic change; and (7) ready tie-in with computer services or other data bases.

There are, however, several disadvantages that must be addressed. These include (1) varied accuracy dependent upon the skill of the user; (2) difficulty in accurate color interpretation due to weak reactions; (3) possibility of media carryover between compartments; (4) inadequacy of test system with some organisms; and (5) occasional need for additional testing. To minimize the potential problems, it is important that the user carefully follow the manufacturer's directions as to the preparation of the inoculum, the incubation times, and the test interpretations.

TYPES OF SYSTEMS

A number of modifications of conventional testing systems have been developed to facilitate the rapidity of bacterial identification. These include the more accurate viewing of organisms, use of smaller and more rapid inoculations, a decrease in incubation time, the automation of the entire process, and the utilization of more systematized identification systems.

Modified Conventional Systems

One of the quickest ways to determine the identity of an organism is microscopic examination. As we have seen, however, it is very difficult to determine the identity of an organism microscopically. If we add an immunological reagent and marker system it is possible to identify microorganisms quickly and accurately. Examples include the fluorescent staining systems for *Chlamydia, Legionella, Neisseria gonorrhoeae, Treponema pallidum,* and several fungal and parasitic organisms.

Another rapid system is the utilization of various agglutination tests. These tend to be less sensitive than the conventional identification methods and are best used as screening systems. Examples include the rapid screening systems for the hemolytic streptococci, *Staphylococcus aureus,* several of the organisms that cause meningitis, *Legionella,* and *Clostridium difficile.*

Conventional testing systems have also been modified to provide more rapid results. Small volumes of media are inoculated with heavy suspensions of the organism to be identified, and paper disks are impregnated with test reagents to provide rapid test determination. Tests that have been modified in this manner include nitrate reduction, urease production, and decarboxylations.

Multitest Systems

Multitest systems are conventional tests packaged in such a manner as to enable a single inoculation of several media or the easy inoculation of several media packaged in small volumes. These provide specific patterns of test results that can be easily compared with the expected results of known organisms. Many of these systems require overnight incubation, but some will give reliable results in as little as 4 to 6 hours. Two of the most commonly used systems for the identification of enteric pathogens are the Enterotube® II (Becton-Dickinson Microbiology Systems, Inc.) and API20E® (Analytab Products, Inc.).

Some of the more recently developed test systems have been adapted so as to not require overnight incubation. These depend upon the detection of preformed enzymes present in a heavy suspension of the organism, giving results following a 4–hour incubation. The Micro-ID system uses several of these enzyme detection systems. Additional 4–hour systems utilize chromogenic substrates that give more rapid color changes; therefore, a rapid reading is possible. An example of this type of system is the RapID STR for identification of streptococci and RapID SS/U for urinary tract bacteria.

Automated Systems

A number of automated identification systems have been developed in recent years. These utilize growth or changes in pH indicators to show reactions. Following an incubation period of 3 to 6 hours, light-scatter photometry is used to determine test results and the identity of the organism. Two of the widely used automated systems are the Vitek® automated system and the MicroScan® automated system.

The Vitek automated system consists of a card containing 30 biochemical tests and an incubator/reader. A number of different cards are available, including ones for gram-negative identification (GNI) and gram-positive identification (GPI) as well as ones for determining antibiotic susceptibility. The cards are placed in the incubator/reader where they are scanned hourly until an identification can be determined. Results are usually read in 4 to 6 hours.

The MicroScan automated system uses a panel containing wells for 26 biochemical tests and 60 wells for antibiotic susceptibility testing. The wells are inoculated with a bacterial suspension and placed in the incubator/reader. They are scanned periodically by the instrument until sufficient results are read to determine the identification of the organism.

ENTEROTUBE II MULTITEST SYSTEM

The Enterotube II consists of a single tube with 12 compartments, each containing a different medium, and a self-contained inoculating needle. It was developed to distinguish among the various Enterobacteriaceae. The needle is touched on an isolated colony and drawn through the tube to inoculate the media. Following incubation, 15 biochemical determinations can be made based on colorimetric changes. The varied positive and negative test results are converted into a numerical ID value, which is either looked up in a code book or determined by the ENCISE computer identification system. The tests included in the Enterotube II system, and possible results are shown in Table 20.1. This was one of the first multitest systems developed and it is still one of the easiest to inoculate and read.

LABORATORY OBJECTIVES

In this exercise phenotypic characteristics of bacterial isolates are used to develop a phenotypic "profile," or numerical identification, for unknown organisms. The profile is then matched with a list of known profiles to identify the organism. To understand the principles demonstrated, you should

- Compare the 'profile' method used in this exercise with the 'dichotomous key' method used in Exercise 18.

- Understand the use of the Enterotube system for the identification of enteric pathogens.
- Understand why a single phenotypic charac-teristic will have a much greater impact in a dichotomous key than it would in a profile analysis key.

MATERIALS NEEDED FOR THIS LAB

1. Twenty-four to 48-hour streak plate of one of the following organisms. The colonies should be well isolated:

   Escherichia coli
   Enterobacter aerogenes
   Proteus vulgaris
   Salmonella typhimurium *
   Shigella sp. *
   Vibrio parahemolyticus *

   *Indicates pathogenic organisms. They may be presented as a demonstration
2. Enterotube II.
3. Test reagents:
   a.   Kovac's reagent
   b.   20% KOH containing 5% alpha-naphthol.
4. Syringe and needle (optional).
5. Identification code manual for Enterotube II.

LABORATORY PROCEDURE

1. Label the Enterotube II with your name and the culture identification. Obtain a streak plate of the organism to be tested.
2. Remove both caps from the Enterotube II. One end of the wire is straight and is used to pick up the inoculum; the bent end of the wire is the handle. Holding the Enterotube II, touch the needle end to an isolated colony. Be careful not to touch the agar with the needle.
3. Inoculate the Enterotube II by holding the bent end of the wire and twisting. Draw the wire through all 12 compartments using a rotating motion. DO NOT remove the wire from the last compartment.

4. Re-insert the needle through the compartment until the notch on the wire is aligned with the opening of the tube (the tip should be in the citrate compartment). Break the needle where it is notched by bending it. The portion of the needle remaining in the tube will maintain the anaerobic conditions necessary for fermen-tation, gas production, and decarboxylation.
5. Punch holes with the broken-off wire through the plastic film covering the air inlets of the last eight compartments (adonitol through citrate) to provide aerobic conditions. Replace the caps on both ends of the tube.
6. Incubate the tube by placing it on its flat surface in a $37^{\circ}C$ incubator for 24 hours.
7. Read and record all of the reactions on the Laboratory Report Form. Be sure to record all results *before* performing the indole and Voges-Proskauer tests. Indicate each positive reaction by circling the number appearing below the appropriate compartment of the Enterotube II report form.

   Indole test: Make a small hole in the plastic film covering the H_2S/Indole compartment. Add 1 to 2 drops of Kovac's reagent. An alternative method would be to inject the Kovac's reagent using a syringe and needle. A positive test is indicated by the development of a red color within 10 seconds.

   Voges-Proskauer (V–P) test: Add 2 drops of 20% KOH containing 5% alpha-naphthol to the V-P compartment. A positive test is indicated by development of a red color within 20 minutes.
8. Add the circled numbers within each bracketed section. Enter the sum in the space provided. Note that the V-P test is only used as a confirmatory test. Read the resultant five digit number and locate it in the code book.
9. Dispose of the Enterotube II as indicated by your instructor.

Table 20.1   Enterotube Biochemical Reactions

TEST CODE	TEST	INITIAL COLOR	FINAL COLOR	INTERPRETATION OF POSITIVE TEST RESULTS
GLU	Glucose fermentation	Red	Yellow	Positive for acid production from glucose
GAS	Glucose fermentation			Positive for gas production from glucose
LYS	Lysine decarboxylase	Yellow	Purple	Positive for the removal of carboxyl group from amino acid lysine
ORN	Ornithine decarboxlylase	Yellow	Purple	Positive for the removal of carboxyl group from amino acid ornithing
H$_2$S	Hydrogen sulfide production		Black precipitate	Positive for the production of hydrogen sulfide which reacts with ferrous ions, forming a black precipitate
IND	Indole production	Beige	Red	Kovac's reagent added to H$_2$S/IND compartment turns red in the presence of indole
ADON	Adonitol fermentation	Red	Yellow	Positive for acid production from adonitol
LAC	Lactose fermentation	Red	Yellow	Positive for acid production from lactose
ARAB	Arabinose fermentation	Red	Yellow	Positive for acid production from arabinose
SORB	Sorbitol fermentation	Red	Yellow	Positive for acid production from sorbitol
VP	Voges-Proskauer	Beige	Red	Voges-Proskauer reagents detect production acetoin
DUL	Dulcitol fermentation	Green	Yellow	Positive for acid production from dulcitol
PA	Phenylalanine utilization		Black precipitate	Positive for the production of phenylpyruvic acid from phenylalanine which then combines with iron salts resulting in black precipitate
UREA	Urea hydrolysis	Yellow	Pink	Production of ammonia from urea
CIT	Citrate utilization	Green	Blue	Citric acid used as carbon source

LABORATORY REPORT FORM
EXERCISE 20
RAPID DIAGNOSTIC TESTS: PROFILE ANALYSIS

What is the purpose of this exercise?

Circle the number corresponding to each positive reaction below the appropriate compartment. Then determine the five–digit code.

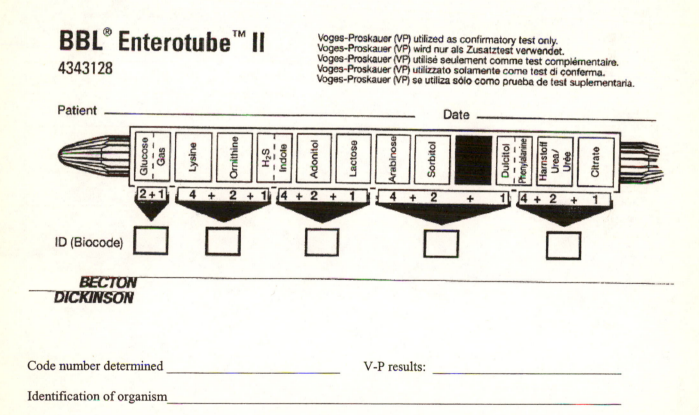

BBL® Enterotube™ II

4343128

Voges-Proskauer (VP) utilized as confirmatory test only.
Voges-Proskauer (VP) wird nur als Zusatztest verwendet.
Voges-Proskauer (VP) utilisé seulement comme test complémentaire.
Voges-Proskauer (VP) utilizzato solamente come test di conferma.
Voges-Proskauer (VP) se utiliza sólo como prueba de test suplementaria.

Patient _____   Date _____

Glucose/Gas | Lysine | Ornithine | H₂S/Indole | Adonitol | Lactose | Arabinose | Sorbitol | | Dulcitol/Phenylalanine | Harnstoff Urea/Urée | Citrate

ID (Biocode)

BECTON DICKINSON

Code number determined _____   V-P results: _____

Identification of organism _____

QUESTIONS

1.   How is it possible for the same species to have two or more identification codes with the Enterotube II?

2.   Why should the first digit in the Enterotube II code always be equal to or greater than 2?

3.   Can the Enterotube II be used on any unknown organism? Why or why not?

MICROBIAL POPULATION COUNTS: MICROSCOPIC COUNTING METHODS

We have used the microscope to determine the size, shape, and arrangement of microorganisms as well as specific cell structures. In this exercise we will learn how to use the microscope to count the number of cells in a population.

There are two general methods of determining the size of cellular populations. One method relies on the actual counting of the cells using specially prepared microscope slides, counting chambers, or electronic particle counters. In the second method, the cells in the population are allowed to grow, and the resultant colonies are counted or some other biological activity (such as increased protein, carbon dioxide production, or acid production) is measured. In Exercise 22 you will determine the size of a bacterial population by the plate–count method, but in this exercise, you will use direct microscope counting methods to actually count yeast and bacterial cells.

BACKGROUND

When more than one method is available to determine what appears to be the same data (e.g., the size of the population), you must often decide which of the methods is best. The answer to this question frequently lies in an understanding of the limitations of the methods and of what each of those methods actually measures. There certainly are many circumstances where plate counting will give all the information you need, but just as certainly there are circumstances where microscopic counting methods provide the quickest and best route to your data. You must choose the method that gives you the most useful data. Microscopic counting methods often provide information that is just not possible to obtain in any other way. Some examples:

- A microscope count includes all cells in the population, whereas a plate count includes only those cells that are capable of growing under the environmental conditions that were used in the experiment—medium, temperature, pH, oxygen content, and so on. For example, anaerobic bacteria will be unable to grow using any of the methods that are not strictly anaerobic. Similarly, if you were to use strict anaerobic methods, the aerobic organisms would be inhibited.

- A plate count cannot, of course, determine the number of nonviable cells in the suspension. For example, if you complete plate counts on a milk sample before and after pasteurization, you should get very different results. A comparison of the counts obtained by microscope and plate-count methodology would yield some information about the quality of the milk before pasteurization and about the effectiveness of the pasteurization method.

- A microscope count will sometimes be the only way of determining the size of the population or of measuring the number of cells in a given volume. A blood count is a good example.

DIRECT MICROSCOPE COUNTING PROCEDURES

A. COUNTING CHAMBERS

A *counting chamber* is a specially constructed glass slide that has a counting grid of known dimensions etched on its surface. A coverslip is supported a known distance above the etched surface. Examine Figure 21.1. Notice the location of the counting chamber and the counting grid.

Figure 21.2 shows, in a composite diagram, a counting chamber grid with Neubauer ruling, and how it might look if you were counting yeast cells or bacteria. The square in the center of the grid, bounded by the double lines, is 1 mm^2. The volume of the fluid in the area bounded by the grid can be calculated because the dimensions of the grid and the space above it are known. Think of the counting area as a chamber bounded on the top by the coverslip, on the bottom by the chamber itself, and on the sides by the dimensions of the counting grid.

The center square, bounded by the double lines, has an area of exactly 1 mm^2. This center square is subdivided into 25 smaller squares, each of which has sides 0.20 mm (or 200 µm) long. They are bounded by heavy lines and are 0.04 mm^2 (1/25 mm^2). The smallest squares (bound by thin lines) have sides that are 0.05 mm (50 µm) long and which are 0.0025 mm^2 (1/400 mm^2).

The grid to be used for counting will depend upon the number and size of the cells in the fluid being counted. For example, if you were counting bacteria, you would use the smaller squares (bounded by the thin single lines), but if you were counting yeast cells, the larger squares (bounded by the heavy lines) would be used. Different magnifications may also be used when counting the cells.

In practice, you would count the cells in several squares (at least five), calculating the average number of cells in each square; you would then multiply by the appropriate numbers to obtain the number of cells per milliliter (ml). You need to know the volume of the fluid being counted in the chamber (multiply the area of the grid by the depth of fluid over the grid) and the dilution of the fluid used to fill the counting chamber.

Typically, the counting-chamber grid used to count bacteria or yeast cells is the central, 1 mm^2 grid. The depth of the space above the grid in a Neubauer chamber is 0.10 mm. The volume of a chamber with a depth of 0.10 mm would therefore be 0.10 mm^3 or 0.10 ml. With the Petroff-Hauser chamber, the depth is only 0.02 mm, giving a resultant volume of 0.02 mm^3 or 0.02 ml. In making the actual count, you should ignore the cells touching the upper and left borders and count all the cells that touch the lower and right margins (even if most of the cell appears to lie on the wrong side of the line).

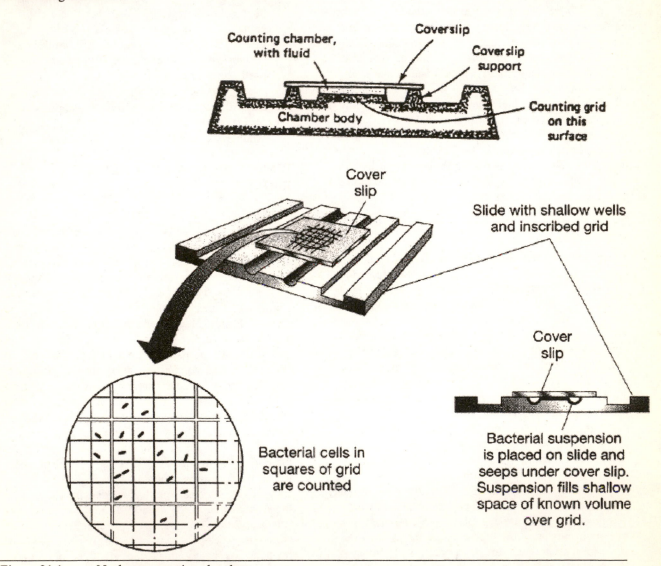

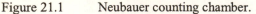

Figure 21.1   Neubauer counting chamber.

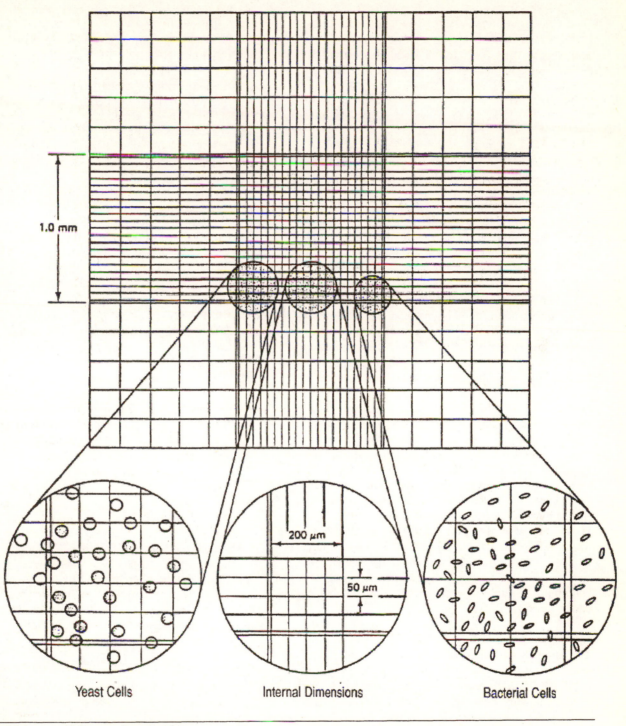

1.0 mm

200 µm

50 µm

Yeast Cells   Internal Dimensions   Bacterial Cells

Figure 21.2   Cell-counting chamber.

In Figure 21.2, there are 17 countable (shaded) yeast cells in the section of the counting grid shown (five are touching the upper and/or left margin). To determine the number of yeast cells in the sample, you would need to multiply the average number counted per square (in this case we will use 17) by the number of squares in the grid (25). This will give you the number of cells in the volume of the counting chamber (0.10 mm^3 or 0.10 ml if you are using a Neubauer chamber).

17 cells/square X 25 squares/grid =
425 cells/0.10 ml

This would then need to be multiplied by 10 to determine the number of cells in 1 ml of the **diluted** sample.

424 cells/0.10 ml X 10 =
4250 cells/1.0 ml

To determine the total number of cells in your original sample, you would have to then multiply by your dilution factor.

How would you calculate the number of bacteria/1.0 ml of **diluted** sample using the grid shown in Figure 21.2? Assume that there is an average count of 11 cells per square. Did you get 44,000 bacteria/1.0 ml of **diluted** sample?

B. DIRECT MICROSCOPE COUNT

In the direct microscope count, a known volume of the suspension is smeared on a slide over a known surface area, usually 1 cm^2. After the suspension dries, the cells can be stained and counted. This method is commonly used to count bacteria in milk and is often referred to as the **Breed count**. In this procedure, 0.01 ml of milk is spread over an area of exactly 1 cm^2. The number of bacteria in several randomly chosen fields of view are counted and the mean number of bacteria per field calculated. If you know the area of the field of view (see Appendix B), you can calculate the number of bacteria in the square centimeter over which the 0.01 ml of suspension was spread by dividing 1.0 cm^2 by the area of the field of view. This will give you the number of fields/cm^2. Multiply the number of fields/cm^2 by the average number of bacteria/field to obtain the number of bacteria in 0.01 ml of the suspension you are counting. Multiply by 100 to obtain the number of bacteria per milliliter.

LABORATORY OBJECTIVES

The microscope can be used to count the number of cells in a suspension. To do this, you must:

- Understand the use of counting chambers and the relationship between the surface area of a grid and the volume of the space above it.
- Understand how the number of cells in a volume of fluid can be determined by the Breed-count procedure.
- Understand that the direct counting method will often, but not always, provide data that cannot be obtained by the plate–count method.

- Appreciate that most experimental methods often have limitations that preclude their use in all instances.

MATERIALS NEEDED FOR THIS LAB
A. COUNTING CHAMBERS
1. Cell-counting chamber (either Neubauer or Petroff-Hauser chambers). Be sure to note which is being used, and therefore the correct measurements of the grid, the number of subdivisions, and the depth of the counting chamber.
2. Coverslip to be used with the counting chamber.
3. Pipettes calibrated to accurately deliver 0.01 ml.
4. Suspension of baker's yeast.
5. A 24-hour heat–inactivated broth culture of *nonmotile* bacteria such as *Bacillus megaterium*.
6. Test tubes containing 9.0 ml of water.
7. Methylene blue stain.
8. Crystal violet stain.

B. DIRECT MICROSCOPIC COUNT
1. Breed-counting slides. If Breed slides are not available, you can use a glass slide as indicated in the procedures.
2. Pipettes calibrated to accurately deliver 0.01 ml.
3. A 24-hour heat–inactivated broth culture of *nonmotile* bacteria such as *Bacillus megaterium*.
4. Milk that is unpasteurized or pasteurized milk that has been kept overnight at room temperature.
5. Test tubes containing 9.0 ml of water.
6. Xylol.
7. 95% alcohol.
8. Methylene blue stain.

LABORATORY PROCEDURE

Obtain a counting chamber and the coverslip that is to be used with it. (Why is this coverslip so much thicker than other coverslips we have used?) Using distilled water, carefully clean the counting surface and the coverslip.

YEAST SUSPENSION
1. Prepare a dilution of yeast suspension by pipetting 1.0 ml of a well–mixed yeast suspension into a tube containing 9.0 ml of water. What is the dilution of your yeast suspension? Add several drops of methylene blue stain to the suspension.
2. Draw a small amount of the diluted yeast suspension into a pipette. Allow some of the suspension to form a bead on the pipette tip. Touch the tip to the edge of the counting chamber (some counting chambers have a

triangular groove cut into their surface for the pipette tip).

3. The droplet will be drawn into the chamber by capillary action and should just fill the chamber. If there is too much fluid in the chamber, it will spill over into the grooves. Any excess fluid should be removed with absorbent paper.

4. Allow the counting chamber to stand for a few moments.

5. Gently place the counting chamber on the microscope stage, with the counting chamber centered over the light source. Locate the grid with the low power of the microscope and position it so that you can count the number of cells in one of the large squares.

6. Change magnification as needed. Count the number of cells in at least 10 squares and calculate the mean number of cells per square.

7. Complete the necessary calculations to determine the number of yeast cells per mililiter (ml) of original suspension.

8. Record your results on the Laboratory Report Form.

BACTERIAL SUSPENSION

1. Mix one loopful of crystal violet with 1.0 ml of a 24-hour broth culture.

2. Follow steps 2 through 7 as indicated for the yeast suspension.

3. Record these results on the Laboratory Report Form.

DIRECT MICROSCOPIC COUNT

Obtain a Breed-counting slide. It will contain from one to five marked staining areas. Each area is 1 cm^2. If a Breed slide is not available, you can approximate one by marking off 1–cm squares with a marker. Use a centimeter ruler to draw a square centimeter on a piece of paper. Use that to outline the square centimeter on the slide.

BACTERIAL SUSPENSION

1. Use a pipette to accurately transfer 0.01 ml of bacterial suspension to one of the squares on the slide. If the slide is clean, the drop will spread evenly over the glass and fill the marked-off area. If necessary, use a loop to spread the fluid.

2. Make both a 1:10 and 1:100 dilution of the bacterial suspension and repeat step 1, applying the suspensions to different staining squares.

3. Allow the smears to air–dry. Heat–fix the smears over a boiling water bath for a few minutes.

4. Stain the smears with methylene blue for 30 seconds.

5. Rinse briefly with distilled water.

6. Allow the slides to air–dry before attempting to count the bacteria.

7. Using the oil-immersion objective, count the number of bacteria in each of at least 10 fields of view. Calculate the mean number of bacteria per field.

8. Calculate the number of bacteria per milliliter of original suspension.

9. Record your results on the Laboratory Report Form.

MILK SAMPLE

1. Use a pipette to accurately transfer 0.01 ml of milk to one of the squares on the slide. If the slide is clean, the drop will spread evenly over the glass and fill the marked-off area. If necessary, use a loop to spread the fluid.

2. Allow the smear to air–dry. Heat–fix the smear over a boiling water bath for a few minutes.

3. Flood the slide with xylol and then rinse gently with 95% alcohol. The xylol removes the fat in milk, and the alcohol washes any residual xylol from the slide.

4. Stain the smear with methylene blue for 30 seconds.

5. Decolorize gently with alcohol until the smear appears light blue. Rinse briefly with distilled water.

6. Allow the slide to air–dry before attempting to count the bacteria.

7. Using the oil-immersion objective, count the number of bacteria in each of at least 10 fields of view. Calculate the mean number of bacteria per field.

8. Calculate the number of bacteria per milliliter of milk.

9. Record your results on the Laboratory Report Form.

NOTES

Name _____

Section _____

LABORATORY REPORT FORM
EXERCISE 21
MICROBIAL POPULATION COUNTS:
MICROSCOPIC COUNTING METHODS

What is the purpose of this exercise?

A.   COUNTING CHAMBERS

Record the following information about the counting chamber you used:

Type of counting chamber:
    (Neubauer or Petroff-Hauser)   _____

Length of one side of the center grid   _____

Area of center grid   _____

Depth of counting chamber   _____

Volume of space over center grid   _____

Length of side of medium squares   _____

Number of medium squares/mm^2 _____

Length of side of smallest squares   _____

Number of smallest squares/mm^2 _____

YEAST SUSPENSION

Number of cells per square in each of 10 fields   ____ ____ ____ ____ ____

    ____ ____ ____ ____ ____

Mean number of cells per square: _____

Calculate the number of cells/ml of your original yeast suspension. Be sure to show your work.

BACTERIAL SUSPENSION

Number of cells per square in each of 10 fields   ____   ____   ____   ____   ____

                                                                        ____   ____   ____   ____   ____

Mean number of cells per square   ____

Calculate the number of cells/ml of your original bacterial suspension. Be sure to show your work.

B.   DIRECT MICROSCOPIC COUNT

Determine the area of the oil-immersion field of your microscope _____

BACTERIAL SUSPENSION

Dilution counted   _____

Number of cells per square in each of 10 fields   ____   ____   ____   ____   ____

                                                                        ____   ____   ____   ____   ____

Mean number of cells per field   ____

Calculate the number of cells/ml of your original bacterial suspension. Show all your work.

MILK SAMPLE

Number of cells per square in each of 10 fields   _____ _____ _____ _____ _____

_____ _____ _____ _____ _____

Mean number of cells per field   _____

Calculate the number of cells/ml of your original milk sample. Show all your work.

QUESTIONS

1.   List three circumstances where a direct microscope count would be preferred over a plate count.

2.   If you made an error in counting your cells that resulted in a mean count of one additional cell, how much error (in other words, how many more cells) would your results include? Use the figures for your bacterial suspension results for comparison.

3.   The Neubauer counting chamber is used to determine the exact number of bacteria present in a urine sample. The sample has been diluted 1:200, and an average count of 19 bacteria/field has been obtained. What would be the number of bacteria/ml of urine?

4.   What would be the significance of the above urine count if the urine sample had been allowed to sit at room temperature for several hours following collection before the count was performed?

MICROBIAL POPULATION COUNTS: VIABLE CELL COUNTS

The size of a bacterial population can be estimated by measuring some biological activity of the population or by plating the cells on a suitable medium and allowing them to grow into visible colonies. The term **viable cell count** refers to the counting of cells by plating them on nutrient medium and counting the colonies that develop. Often the results are referred to as **plate counts** or **colony counts** and the results reported as **colony-forming units (cfu's)**.

Colony counts have some distinct advantages when compared to direct microscopic counts. The colony count is significantly more sensitive than the direct count. Some microbiologists claim that there must be more than 50,000 bacteria in each milliliter before one can reliably and accurately count them on a microscope slide. Conversely, plate-counting procedures, particularly when membrane-filtration techniques are used, can detect very few cfu's in relatively large volumes of liquid. When manually counting colonies, it is important that there be at least 30 colonies on the plate before the count is considered to be a representative sample. Why?

BACKGROUND

SOLUTIONS, SUSPENSIONS, AND DILUTIONS

A word or two on terminology. A solution results when one chemical (the **solute**) is dissolved in another, usually liquid, chemical (the **solvent**). A good example would be a solution of sodium chloride (salt) in water. The salt, being soluble in water, is the solute, while the water is the solvent. The salt is soluble in water. A **suspension** results when one of the components is not soluble in the liquid. If you added sand or dirt to water, you would have a suspension of dirt or sand in water. Usually, but not always, solutions are clear whereas suspensions are turbid.

It is often necessary to further dilute suspensions and solutions. The dilutions are expressed as relative volumes, such as 1 to 10 or 1 to 100 (written as 1:10 or 1:100). To make a 1:10 dilution, you would add 1.0 ml of the suspension or solution to 9.0 ml of diluents. Using the salt solution as an example, if you transferred 1.0 ml of a 1% solution to 9.0 ml of water, you would have made a 1:10 dilution of the original solution

Most bacterial and viral samples can be treated as suspensions. If the sample is liquid, we determine the number of bacteria or viruses suspended in the liquid, diluting it as needed. If it is solid, you must first make a suspension (typically 10% in water) by suspending the sample in a liquid. When these samples are counted, the concentration of the initial suspension must be calculated into the dilution factors. For example, a 1.0% suspension has already been diluted 1:100.

SERIAL DILUTIONS

One of the most important quantitative techniques routinely used in clinical microbiology laboratories is the *serial dilution* procedure. When a sample has been diluted, each of the dilutions can be tested to determine which of them give positive test results. Alternatively, the number of bacteria or viruses in each of the diluted samples can be counted after they grow on a suitable medium.

When the number of bacteria or viruses in a sample must be counted, the report usually indicates the number of bacteria or viruses in each milliliter of the original sample. The report might read "15,000 bacteria/ml" or "150,000 phage/ml." If one encounters very large numbers of bacteria, the logarithmic number is often used instead of the arithmetic number. This, of course, does not change the results at all, but it does make things somewhat more convenient. The actual number of bacteria or viruses is determined by multiplying the number of colonies (or plaques) observed on the plate times the dilution (assuming 1.0–ml aliquots were plated out).

Number of bacteria/ml =
number of colonies × dilution

A serial dilution is typically a stepwise dilution sequence made by transferring aliquots from one tube of diluents to another. The aliquots are usually the same volume throughout the sequence. If a 1:10 dilution sequence is used the process is referred to as a tenfold serial dilution; if a 1:2 dilution sequence is used, it is a twofold serial dilution. The number (ten, two, etc.) is the dilution accomplished at each step, a tenfold serial dilution being a series of 1:10 dilutions. The following table has some examples:

Serial Dilution		Volumes Used (Aliquot transferred into)
Tenfold	1:10	1.0 ml into 9.0 ml
Fivefold	1:5	1.0 ml into 4.0 ml
Twofold	1:2	1.0 ml into 1.0 ml

Because a serial dilution is a series of identical, stepwise dilutions, the final dilution, or the dilution of any tube in the sequence, can be easily determined. The highest dilution is always in the last tube, whereas the lowest dilution is always in the first tube. Using zero (0) to indicate the undiluted sample is sometimes helpful in keeping track of the dilution steps. When bacterial or viral suspensions with very large numbers of organisms are being diluted, the undiluted sample is frequently not counted. When antibody titer in serum is to be determined, or when the expected numbers are very low, the undiluted sample is often counted or assayed.

POUR-PLATE COUNTING OF BACTERIA

If you added 1.0 ml of a bacterial suspension to melted agar and then poured the agar into a petri dish, the bacteria would be immobilized in the agar when it solidified. After a suitable incubation time, the colonies would become visible and you could count them (see Exercise 10). Figure 22.1 shows a typical tenfold serial dilution/pour plate-counting protocol.

Although a single colony may contain more than a million bacterial cells, it can be assumed that all of those cells arose from a single cell. The number of colonies is, therefore, an accurate representation of the number of viable cells (cells able to grow on the medium used and under the incubation conditions employed).

If you did one of these counts on each tube in a serial dilution, you would have completed a plate count. Your results might look like those in Table 22.1.

When counting colonies, only those plates that have between 30 and 300 colonies are counted. In the example shown in Table 22.1, the only "countable" plate is plate 5. The reason for counting plates that have between 30 and 300 colonies is that this provides us with the most representative samples. Fewer than 30 colonies is really too few to be representative. More than 300 colonies results in the overcrowding of colonies and poor growth. In addition, it becomes extremely difficult to count those plates accurately!

The results from the plate count are given as the number of colony-forming units per milliliter of the original culture (cfu/ml). To calculate this value, it is necessary to multiply the actual colony count by the dilution factor. For example, for plate 5 in Table

22.1, 38 colonies were counted in a 1:100,000 (10^{-5}) dilution. When the actual count is multiplied by the dilution factor you get a value of 3,800,000 cfu/ml. This can also be represented as 3.8×10^6.

Table 22.1 Results of a Bacterial Plate Count

PLATE NUMBER	DILUTION	LOG 10 DILUTION	COLONY COUNT
1	1:10	-1	TNTC*
2	1:100	-2	TNTC
3	1:1,000	-3	TNTC
4	1:10,000	-4	321
5	1:100,000	-5	38
6	1:1,000,000	-6	5
7	1:10,000,000	-7	0

*Indicates colonies were too numerous to count.

CONSIDERATIONS USED IN PLATE COUNTING

Certain assumptions and considerations have to be taken into account in all plate-counting procedures. These include:

1. Each colony is assumed to be the progeny of a single cell. If we could be assured that each colony did arise from a single cell, then the assumption that the number of colonies equals the number of cells in the suspension could be accepted. In fact, while it is true that some types of bacteria do readily separate into single cells, many do not.

   The variability in the tendency of bacterial cells to remain clumped will introduce some error to your counts unless you are careful to treat each sample the same. For example, you should mix the dilution blanks in a consistent manner (e.g., 20 shakes, 1 minute on a vortex mixer at a given speed, etc.).

   Because of this tendency for organisms to remain clumped, many microbiologists prefer to use the term **colony-forming units per milliliter** (cfu/ml) instead of bacteria per milliliter. They point out that it is the ability of the cells to grow and to exercise their usual biological activity that is of concern. It really does not matter whether the colony (or infectious unit) arose from a single cell or not; what does matter is that it will grow into a colony or may cause an infection.

2. Except for relatively rare and unusual circumstances, plate counts of naturally occurring populations actually select for relatively small segments of these populations. The growth medium that is used in the petri plates as well as the conditions under which the plates are incubated are themselves selective.

   Pure cultures of laboratory strains of bacteria, under some circumstances, represent an example where all cells might be expected to grow when transferred to a petri plate containing an appropriate nutrient medium. Most natural populations of bacteria, such as would be found in the intestinal tract, in water samples, or in soil samples, contain bacteria that are not able to grow in the environmental conditions used for plate counting. Often special media must be used, or other environmental conditions must be manipulated. These include light, pH, anaerobiosis, temperature, special sources of oxidizable carbohydrate, and so on.

3. Frequently the selectivity of the medium or the growth environment can be used to encourage

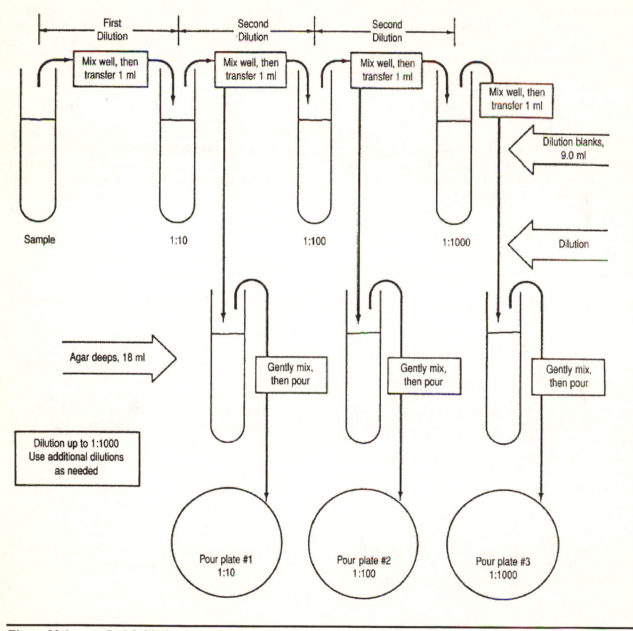

Figure 22.1   Serial-dilution procedure.

the growth of specific kinds of bacteria. For example, we might want to know how many anaerobic bacteria were in the population. In soil microbiology it is important to know how large the cellulose-decomposing segment of the population is. In water, we need to know the number of coliforms before we can determine whether the water is safe to drink. In clinical microbiology, the presence of more than 100,000 bacteria in urine is considered significant, but in most other cases where a normal flora is usually present, we are only interested in certain types of bacteria (beta-hemolytic, lactose nonfermenters, etc.).

4. It is often necessary to perform both a direct microscopic count and a plate count on a sample. What would you conclude from a milk sample that had a plate count of 1,000 cfu/ml and a direct count of 25,000 bacteria/ml?

SOURCES OF ERROR IN SERIAL DILUTIONS

Two common sources of error introduced in the serial-dilution procedure are pipetting errors and errors introduced by failing to mix the diluted samples sufficiently. The pipetting errors tend to resolve themselves with practice. As you gain experience, the reproducibility of your results will improve. If you remember a few rules about pipetting, your errors will be markedly reduced.

1. Always determine the type of pipette you are using and then use it correctly. A measuring pipette is designed to accurately deliver the volume indicated between two marks on the pipette. The graduations on a measuring pipette do not extend to the tip. A serological pipette is designed to deliver the stated volume by allowing the fluid to flow out of the pipette under the force of gravity. The graduations on this type of pipette always extend to the tip.
2. Always hold the pipette in the vertical position. This way the meniscus can always be read more accurately.
3. Always use the bottom of the meniscus to measure volume. Because of surface–tension properties present in aqueous solutions, a convex meniscus forms in the pipette. By always using the bottom of the meniscus, it is possible to maintain accurate measuring with pipettes.

Mixing errors can be avoided by taking care to always mix the sample carefully. You should mix all the tubes the same way—for example, by vortex mixing for 1 minute, or by shaking 20 times. It really does not matter how you mix the sample, as long as

you are consistent and careful to ensure complete mixing.

1. Always mix the dilutions carefully.
2. If a vortex mixer is available, vortex each tube for the same amount of time.
3. If you use bottles or screw-cap tubes, shake each bottle or tube the same number of times.
4. The key to accurate dilutions is consistency; mix and pipette each dilution the same way.

The serial-dilution protocol will use 1.0-ml aliquots for plating. It is sometimes convenient to plate out 0.1 ml and make the appropriate mathematical changes in dilution calculation. Using the smaller volume allows you to stretch out your dilution sequence, but it also makes pipetting errors proportionately larger.

SOME REMINDERS

Plate counts usually require that the sample be diluted before plating. Study the diagram in Figure 22.1. The reason for the dilution of the sample is to ensure that you will have plates with between 30 and 300 colonies on them. As a starting point, a 24-hour culture of *E. coli* will contain a cell population of between 1 million and 100 million cells per milliliter.

SOME APPLICATIONS

While it is often important to know the size of a bacterial population, it is sometimes even more important (and interesting) to learn how that population changes. For example, what would be the effect of changing the nutrients in a medium? How would you go about testing which medium produced better growth—nutrient broth or brain-heart infusion broth?

What results would you expect to obtain if you did a plate count on a broth culture every 2 or 3 hours for about 24 hours? Would you be able to demonstrate the typical growth curve? Why not try it?

How would you evaluate the quality of water or milk? Most foods must meet standards that include limits on the numbers of bacteria per gram or milliliter. Often they will specify limits that include total bacteria as well as total coliforms. How would you modify this exercise to detect coliforms? (Look at Exercise 39 for possible suggestions.)

LABORATORY OBJECTIVES

Viable cell counts determine the number of living cells in a population. Although this information may be very important, there are some limitations on the procedure. In this exercise you should:

- understand the relationship between the serial dilution and the plating of the sample;
- be able to list several of the limitations to the plate–count procedure;
- be able to discuss how plate counting and direct counting can often provide different, yet complementary, information about the population being studied.

MATERIALS NEEDED FOR THIS LAB

1. 9.0 ml water or saline (0.9% NaCl) dilution tubes: 7 tubes if a tenfold serial dilution to 1:10,000,000 is to be performed.
2. Nutrient agar deeps—1 for each dilution being plated.
3. Sterile 1.0-ml pipettes.
4. Sterile Petri plates.
5. Sample to be plated. Any available source such as water, soil, milk or a food product can be used. An 18- to 24-hour broth culture of *Staphylococcus epidermidis* or *Escherichia coli* could also be used.

LABORATORY PROCEDURE

1. Obtain and label all the dilution blanks, nutrient agar deeps, pipettes, and petri plates you will need to complete your dilutions and platings. Remember to label the petri plates so that your labels will not obstruct your view through the bottom of the plate.
2. When melted, the agar deeps should be maintained in a water bath (approximately 50°C)

to prevent the medium from solidifying before you use it.

3. Arrange your material so that everything is within reach.
4. Decide which of the dilutions will be plated. If the culture or sample you are working with has about 10,000,000 cells/ml, you should plate the 1:100,000 through 1:10,000,000 dilutions as a minimum.
5. Following the diagram in Figure 22.1, complete your serial dilutions and plating simultaneously. Use the same pipette to transfer 1.0 ml from one dilution blank to the melted agar and to the next dilution blank.
6. Allow the poured agar to solidify before moving the plates.
7. Incubate the plates for 24 hours at 37°C.
8. Count the number of colonies on each plate. If the medium is transparent, you should count the colonies through the bottom of the plate. It is sometimes helpful to mark each colony with a felt-tip pen as you count it.
9. Use any plates that have between 30 and 300 colonies to calculate the number of colony–forming units (cfu) per milliliter in the original sample. Do not forget to take into account your dilutions. You do not need to allow for the dilution of the sample when it is added to medium. Why not?
10. Record all results and calculations on the Laboratory Report Form.

NOTES_____

Name _____

Section _____

LABORATORY REPORT FORM

EXERCISE 22
MICROBIAL POPULATION COUNTS: VIABLE CELL COUNTS

What is the purpose of this exercise?

PLATE–COUNT RESULTS

Report your plate-count results in the following table. Give the actual number of colonies counted at each dilution and calculate the number of cfu/ml in the original sample. Finally, convert the counts to their logarithmic numbers.

OBSERVED DATA		CALCULATED RESULTS	
DILUTION	COLONIES COUNTED	ACTUAL COUNT (cfu/ml)	LOG 10 OF COUNT (cfu/ml)
MEAN VALUES:			

QUESTIONS

1. Why are counts above 300 and below 30 statistically unreliable?

149

2. What factors would contribute to plate counts being selective?   How could their selective nature be minimized?

3. The number of bacteria in a milk sample has been determined by preparing a tenfold serial dilution series. Five dilution tubes were utilized and the final three dilutions were plated.   The first plate was too numerous to count (TNTC), the second had 198 colonies, and the third plate had 23 colonies. Determine the number of bacteria in the original sample (cfu/ml).

PHYSICAL CONTROL METHODS

In Exercises 11 and 12, we examined the physical growth requirements of several organisms. Just as we use our knowledge of an organism's optimal physical growth requirements to guarantee its successful cultivation, we can use this information to inhibit the organism's growth.

BACKGROUND

Each organism has an optimum temperature, osmotic pressure, oxygen level, and pH. In addition, all living cells require the maintenance of adequate water concentrations to survive. Often, these organisms are able to survive for at least limited periods of time in the absence of optimal conditions. Contributing to an organism's survival rate under less than optimal conditions would be the presence of an endospore, the growth phase of the organism, and the nutrients present.

Microbial control methods include **sterilization, disinfection** and **sanitization**. Sterilization is the destruction or removal of all microbial organisms and their products in or on an object. Disinfection, on the other hand, is the destruction, inactivation, or removal of microorganisms likely to cause undesirable effects such as disease or spoilage. (Does a disinfectant sterilize?) To sanitize means simply to reduce the number of organisms present with no guarantee of their destruction.

We often cannot depend solely upon sanitization, but need to be able to guarantee the destruction of certain microorganisms. This might be in the sterilization of bacteriologic media or surgical instruments, the disinfection of the laboratory bench or an operating room, the antisepsis of a patient's skin prior to surgery or following a cut, or the preservation of our food. It requires that we have some knowledge of the organisms that we are most likely dealing with and the conditions under which they can be destroyed. In this exercise we will examine the role of several physical conditions in microbial control.

TEMPERATURE

We saw in Exercise 12 that each organism has an optimum temperature as well as a range in which it survives. Each organism also has a **thermal death time** and **thermal death point** that can be determined. The thermal death time is the shortest period of time required to kill a suspension of a given microorganism at a given temperature and under specified conditions. It is important to note that this must be determined for every species. The thermal death point is the lowest temperature at which a given microorganism is killed in a given time (usually 10 minutes). Knowledge of these values are important in determining the conditions that will be used for the control of microorganisms by heat.

Heat has a number of applications in microbial control. This involves both moist heat (autoclaving, pasteurization, or boiling water disinfection) and dry heart (hot air oven and incineration) methods. No matter which method is used, its ability to control microbial growth is dependent upon its interference with normal enzymatic activity. When enzymes are heated too greatly, they can no longer maintain their proper shape and become inactive, a process known as *denaturation of protein*. Without proper enzyme functioning, cells cannot function and death occurs. The difference in the amount of heat needed and its overall effectiveness will vary dependent upon both the ability of the heat to penetrate the cell and the presence of resistant structures such as endospores.

It should also be noted that enzyme activity slows down as the temperature of an organism decreases. This too can serve to diminish the catalytic capabilities of the enzyme and inhibit the growth of the cell. The lowering of temperature, however, is not a guarantee of cellular death (Can meat taken from the freezer still spoil even though it does not come in contact with new contaminants?)

In this exercise you will determine the thermal death time for a given organism and also examine the effect of freezing temperatures on microbial growth.

OSMOTIC PRESSURE

We also saw in Exercise 12 that each organism has a range of osmotic pressure in which it could survive or actively grow. Hypotonic environments are rarely lethal to microorganisms owing to the presence of a rigid cell wall that resists the flow of too much water and the rupturing of the cell. (What does the term *hypotonic* mean? If you are not sure, go back to Exercise 12 and reread the background material.) *Hypertonic* solutions are far more effective in microbial control. In fact, hypertonic solutions were the first major methods of food preservation. Perhaps the greatest hypertonic environment we can present to an organism is that which occurs in the total absence of water or when drying occurs. Why is this considered hypertonic? What would be the effect on the bacterial cell? How do we use various hypertonic solutions today to preserve food? Remember this can involve the increased concentration of either salt or sugar or the decreased concentration of water.

LABORATORY OBJECTIVES

In this exercise we will be examining the effect that elevated temperature, lowered temperature, salt concentration, and drying have on microbial growth. When performing this exercise you should:

- propose, perform, evaluate, and appraise the results of an experiment that establishes thermal death time.

- examine the relationship between the physical growth requirements of an organism and the physical methods of microbial control.

MATERIALS NEEDED FOR THIS LAB

A. THERMAL DEATH TIME
   1. Twenty–four–hour broth culture of one of the following
      Bacillus subtilis
      Staphylococcus epidermidis
      Escherichia coli
   2. Two TSA plates
   3. One sterile test tube
   4. 100°C water bath
   5. Test tube rack.

B. THE EFFECT OF FREEZING
   1. Twenty–four–hour broth cultures of
      Bacillus subtilis
      Staphylococcus epidermidis
      Escherichia coli
   2. Three sterile test tubes
   3. Two TSA plates
   4. Sterile 5.0 ml pipettes and pipetter.

C. THE EFFECT OF OSMOTIC PRESSURE
   1. Twenty–four–hour broth cultures of
      Bacillus subtilis
      Staphylococcus epidermidis
      Escherichia coli
      Halobacterium salinarium
   2. TSA tall (15 ml) containing 15% NaCl
   3. TSA tall (15 ml) containing 0.5% NaCl
   4. Sterile petri plate
   5. 50°C water bath

D. THE EFFECT OF DRYING
   1. Twenty–four–hour broth cultures of
      Bacillus subtilis
      Staphylococcus epidermidis
      Escherichia coli
   2. Three sterile screw-capped test tubes
   3. Three sterile swabs
   4. Disinfectant
   5. Sterile TSA broth
   6. Sterile 5.0 ml pipettes and pipetter.

LABORATORY PROCEDURES

A. THERMAL DEATH TIME
   Remember, thermal death time is the amount of time required to kill bacteria and their spores at a given temperature. You are going to design an experiment to determine the thermal death time of one of the given organisms using the scientific method. This requires you to:
   a. state the problem;
   b. formulate a testable hypothesis;
   c. make a prediction;
   d. design and perform an experiment, being sure to incorporate adequate controls;
   e. form a conclusion based on experimental results;
   f. re-evaluate experimental design; and
   g. if necessary, formulate another testable hypothesis and test it.

   The following limitations are placed on your experimental design:
   a. The test organism will be the organism assigned by your instructor.
   b. The temperature used for your experiment will be 100°C.
   c. The equipment available will be limited to:
         2 TSA plates
         1 sterile test tube
         100°C water bath
         Test tube rack.

1. Determine your hypothesis. What is it that you are testing?
2. Determine your experimental design. Remember, TSA plates can be divided into varied number of sectors. Be sure that your experiment is designed to answer the initial question! Record your experimental design on the Laboratory Report Form
3. Place the designated culture into the 100°C water bath, removing a loopful at the time intervals present in your experimental design. Streak in the appropriate sector on your TSA plate.

4. Incubate the plates at 37°C until the next laboratory period.
5. Examine your plates. Did you determine the thermal death time for your given organism? What changes would you make if you were to repeat the experiment?

B. THE EFFECT OF FREEZING

1. Obtain a TSA plate. Divide the bottom into three sectors, labeling each with the name of one of your test organisms.
2. Inoculate each sector with one of the broth cultures. (What is the purpose of this plate?)
3. Obtain three sterile test tubes. Label each with the name of one of the organisms.
4. Pipette 3.0 ml of the first broth culture into its labeled sterile tube. Repeat with new sterile pipettes for each organism.
5. Place the tubes into the freezer until the next laboratory period.
6. Remove the cultures from the freezer and allow them to thaw. They may be placed in tepid water to thaw.
7. Observe the plate from day one. Record the presence or absence of growth for each organism on your Laboratory Report Form.
8. Obtain a TSA plate. Divide it into three sectors and label one sector for each organism.
9. Inoculate each sector with the appropriate thawed culture.
10. Incubate at 37°C until the next laboratory period.
11. Following incubation, observe the plate for growth. Record your results on the Laboratory Report Form.

C. THE EFFECT OF OSMOTIC PRESSURE

1. Place a tube of melted TSA tall containing 15% NaCl and a tube of melted TSA tall containing 0.5% NaCl in a 50°C water bath.
2. Put a pencil on your laboratory bench. Place a sterile petri plate so that one edge is resting on the pencil

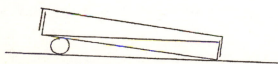

3. Carefully add the TSA containing 15% NaCl so that it does not quite cover the bottom of the plate. Cover the plate with its lid and let it sit on the pencil until the agar sets (approximately 15 minutes).
4. When the bottom wedge of agar has solidified, label the side of the plate with the thickest portion of agar "15%." Place the plate flat on the

lab bench. Carefully pour the TSA tall containing 0.5% NaCl over the surface of the agar wedge until it just flows over the thickest portion of the wedge. Replace the cover on the plate and let it sit until it is completely solidified.
5. After the plate is solid, draw four evenly spaced parallel lines from the 15% NaCl side to the 0.5% NaCl side of the plate. Label each line with the name of one of the test organisms.

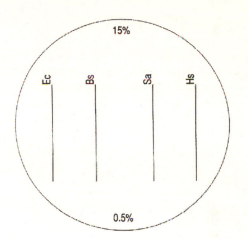

6. Using your broth cultures, inoculate each along its marked line, starting at the low concentration end of the line.
7. Incubate the plate at 37°C until the next laboratory period.
8. Following incubation, record the growth patterns observed on you Laboratory Report Form.

D. THE EFFECT OF DRYING

1. Obtain three sterile screw-capped test tubes, three swabs and the broth cultures. Label one tube for each organism being tested.
2. Dip a sterile swab into a broth culture. Be sure to ring the inside of the tube with the swab so the swab is not too wet.
3. Swab the inside of the labeled screw-capped tube with the culture. Be careful not to leave a large drop in the tube.
4. Repeat steps 2 and 3 with the other organisms.
5. Place all tubes in the 37°C incubator until the next class period.
6. Pipette 3.0 ml of sterile nutrient broth into each of the screw capped tubes.
7. Re-incubate until the next laboratory period.
8. Observe each tube for the presence of growth.
9. Record your results on the Laboratory Report Form.

NOTES

LABORATORY REPORT FORM
EXERCISE 23
PHYSICAL CONTROL METHODS

What is the purpose of this exercise?

A. THERMAL DEATH TIME

1. What is your hypothesis about the thermal death time of your given microorganism?

2. What is your experimental design? Be sure to include adequate controls!

3. Report the results for your experiment.

4. Did you determine the thermal death time for your organism? _____

    If so, what is it?

    If not, what would you vary if you were to repeat the experiment?

B. EFFECT OF FREEZING

1. Complete the following table indicating the presence (+) or absence (−) of growth.

Freezing time: _____

ORGANISM	CONTROL	FOLLOWING FREEZING
Escherichia coli		
Bacillus subtilis		
Staphylococcus epidermidis		

2. What effect did freezing have on the survival of the test organism?

C. OSMOTIC PRESSURE

1. Record your results in the diagram.

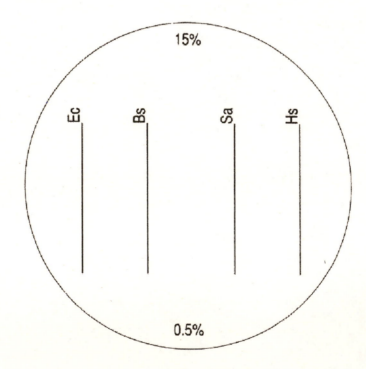

2. Approximately what concentration of NaCl, if any, was inhibitory to the test organisms?

Escherichia coli _____

Bacillus subtilis _____

Staphylococcus epidermidis _____

Halobacterium salinarium _____

D. EFFECT OF DRYING

1. Complete the following table indicating the presence (+) or absence (−) of growth. What was the control for this experiment?

ORGANISM	CONTROL	FOLLOWING DRYING
Escherichia coli		
Bacillus subtilis		
Staphylococcus epidermidis		

2. What effect did drying have on the survival of the test organisms?

QUESTIONS

1. Why would you expect the thermal death time for *Escherichia coli* or *Staphylococcus epidermidis* to differ from that for *Bacillus subtilis*?

2. Why is *Clostridium botulinum* a major concern in the canning of food?

3. Culture strips of *Bacillus subtilis* or *Bacillus stearothermophilus* often serve as quality–control systems for autoclaving. Why are these organisms used?

4. Give examples of how we use each of the following in the preservation of food:

Elevated temperature:

Lowered temperature:

Osmotic pressure:

Dehydration:

ULTRAVIOLET LIGHT AS AN ANTIMICROBIAL AGENT

Ultraviolet light is often used to disinfect or sterilize surfaces of objects that cannot be conveniently sterilized by other methods. It is also used to sterilize air entering and leaving certain types of "clean rooms." Its effectiveness as a bactericidal agent is based on the light absorbing properties of nucleic acids.

CAUTION: Ultraviolet light can damage the retina of the human eye. Under no circumstances should you look into the lamp used in this experiment. Avoid having reflected light shine into your eyes. Avoid direct exposure of the skin with ultraviolet light.

BACKGROUND

THEORETICAL BASIS FOR BACTERICIDAL EFFECTS OF ULTRAVIOLET LIGHT
Sunlight is composed of a continuous spectrum of electromagnetic radiation with varying wavelengths (sSee Figure 24.1). Visible light consists of those wavelengths between 400 and 900 nm (a nanometer or nm $= 10^{-9}$ m). Those wavelengths above the visible spectrum are the infrared rays, microwaves, and radio waves. Those below 400 nm fall into the ultraviolet waves, X-rays, and gamma rays. In this exercise we are most concerned about the

wavelengths in the ultraviolet range (100 to 400 nm). Ultraviolet rays below 200 nm are readily absorbed by air. The ultraviolet rays that exhibit the greatest affect on cells, specifically the DNA of those cells, fall between 200 and 290 nm or the UVC wavelength. Wavelengths between 250 and 260 nm have been found to be the most injurious to cells.

The ultraviolet rays stimulate the formation of bonds between the carbons of adjacent pyrimidines (see Figure 24.2). The resulting **pyrimidine dimers** interfere with the proper pairing of complementary bases, thereby interfering with proper DNA replication by the incorporation of improper nucleotides. This results in mutation or, in severe cases, death of the cell. It should be noted that not all ultraviolet-induced mutations of DNA are lethal or detrimental. Mutations can also have a beneficial or neutral effect on the viability of the affected cell.

Cells have two mechanisms that enable them to reverse the effect of ultraviolet light on the DNA —**photoreactivation** and **dark repair** (see Figure 24.3). Photoreactivation, or light repair, occurs in the presence of light. Light activates an enzyme that breaks the bond in the pyrimidine dimer, returning DNA to its original state.

Dark repair, which does not occur in all bacteria, involves the utilization of several enzymes.

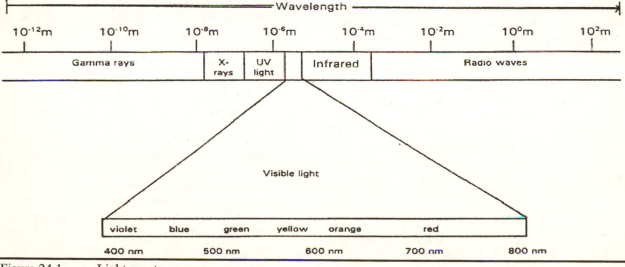

Figure 24.1   Light spectrum.

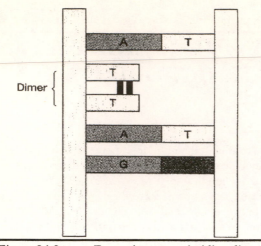

Figure 24.2   Formation on pyrimidine dimer

Endonucleases "cut out" the dimer, DNA polymerase synthesizes new DNA to replace this section, and ligase attaches the new segment to the original strand.

USE OF ULTRAVIOLET LIGHT AS A BACTERICIDAL AGENT

There are limits to the application of ultraviolet light as an antimicrobial agent. Exposure of items does not always result in sterilization. Ultraviolet rays have very poor penetrating abilities and their effect is readily blocked by glass, water, and the presence of other organic material. Pigmented and endospore-forming organisms are more resistant to its effects than are actively growing non-pigmented organisms. (What hypothesis can you form as an explanation for this?) The effectiveness of ultraviolet light is influenced by the distance between the source of the light and the cells. The closer the light is to the cells, the more intense its effect. In addition, there are the problems of both photoreactivation and dark repair, which can reverse the effects of the ultraviolet light.

Another problem that must be addressed before ultraviolet light can be used is its safe application. The effect of ultraviolet light on DNA is not restricted to bacterial cells and bacterial DNA. The same effects occur in any cell undergoing direct exposure to ultraviolet light. (Why is our overexposure to sunlight discouraged?) It is recommended that individuals not be present when ultraviolet lights are being used or that the lights are adequately shielded to protect individuals in the area.

Ultraviolet light has been used for the destruction of airborne organisms in operating rooms, pharmaceutical companies, research laboratories, and food processing areas—but primarily in the absence of people. It has also been used for the sterilization

of heat–labile solutions and disinfection of wastewater.

EXPERIMENTAL DESIGN

This experiment is designed to demonstrate the bactericidal properties of ultraviolet light. To accomplish this, you will suspend bacteria in saline. Bacteria will be exposed to ultraviolet rays and, at given time intervals, a loopful of the irradiated suspension will be transferred to fresh TSA plates. Any bacteria that have survived exposure will produce viable growth after incubation.

LABORATORY OBJECTIVES

In this exercise you will determine the time needed to kill suspensions of bacteria exposed to ultraviolet radiation. To complete this exercise, you will need to

- understand the mechanism by which radiation accomplishes its bactericidal effects;
- explain the environmental limits to the use of ultraviolet radiation;
- understand the reasons why certain components of the environment may protect suspended bacteria from ultraviolet-induced lethal effects;
- discuss commercial and medical applications of ultraviolet light;
- identify two repair mechanisms that enable cells to survive exposure to ultraviolet radiation;
- demonstrate the effect of endospores and pigment on the effectiveness of ultraviolet radiation.

MATERIALS NEEDED FOR THIS LAB
1. Two TSA plates per group
2. Broth cultures of the following:
   Escherichia coli (24 hour)
   Serratia marcescens (24 hour)
   Bacillus subtilis (72 hours)
   Each group will test one of the organisms, as assigned by your instructor.
3. One tube of sterile saline.
4. One sterile petri dish.
5. Sterile pipettes and pipetter.
6. Ultraviolight with a wavelength between 250 and 260 nm located 6" to 9" above the surface.

LABORATORY PROCEDURE
1. Obtain two TSA plates. Divide each plate into four quadrants by marking the bottom of the plate with a glass wax marker.
2. Label the quadrants on the first plate "0," "30 sec," "1 min," and "2 min." The second plate should be labeled "3 min," "5 min," "10 min," and "20 min."

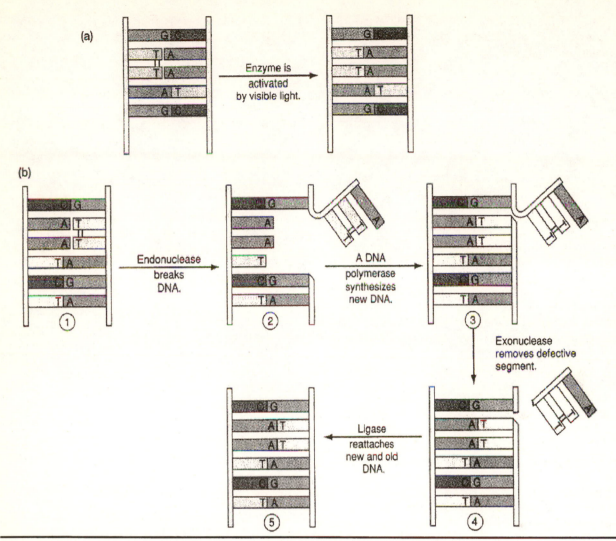

Figure 24.3   Repair of pyrimidine pimers: (a) photoreactivation; (b) dark repair.

3. Each group will be assigned one of the organisms to test.

4. Transfer 1 ml of the assigned culture to the tube of sterile saline. Mix well.

5. Pour about 3 ml of the mixture into a sterile petri plate.

6. Transfer one loopful of the bacterial suspension in the petri plate labeled "0." This will serve as your control.

7. Place the open petri plate (it must be open because the ultraviolet light may be absorbed by the glass or plastic petri plate cover) under the ultraviolet light source. Be sure that the entire plate is exposed to the light. Begin timing immediately after putting the plate under the light source.

8. At the designated times, transfer one loopful of the suspension to the appropriate quadrant of the petri plate.

EXPOSURE TIME:

    30 sec
    1 min
    2 min
    3 min
    5 min
    10 min
    15 min

9. Incubate plates as indicated:
    Escherichia coli, 37°C
    Serratia marcescens, room temperature
    Bacillus subtilis, 37°C.

10. Observe the plates for growth. With *Serratia marcescens* also note the presence of nonpigmented colonies.

11. Record your results on the Laboratoary Report Form.

NOTES

LABORATORY REPORT FORM

EXERCISE 24
ULTRAVIOLET LIGHT AS AN ANTIMICROBIAL AGENT

What is the purpose of this exercise?

1. What was the distance between the ultraviolet light and the petri plate?

2. Complete the following table. Be sure to get the data from others in class for the organisms you did not test. Record the amount of growth as 4+ (heaviest growth), 3+, 2+, 1+ (slight growth) or 0 (no growth).

TIME	*Escherichia coli*	*Serratia marcescens*		*Bacillus subtilis*
		Growth	Pigmentation	
0				
30 sec				
1 min				
2 min				
3 min				
5 min				
10 min				
15 min				

3. Did you observe any variation in pigmentation in *Serratia marcescens*? What variation did you observe and in which plate(s)?

4. How might you explain any pigment variation observed in *Serratia marcescens*?

QUESTIONS

1. Why must the petri plate be left open during the exposure to ultraviolet light?

2. How is DNA damaged by ultraviolet light?

3. What is photoreactivation?

4. What other forms of radiation are lethal to microbes? Are they used as antimicrobial agents?

ASSAY OF ANTIMICROBIAL AGENTS: DISK-DIFFUSION METHODS

How effective are the antiseptics and disinfectants you use? Can you effectively sterilize the counter or tabletop? Can the antiseptic used to cleanse the skin also be used to sterilize instruments? How effective are all of those new antibacterial liquid soaps and household cleaning agents? These and many other questions must be answered if effective sterilization and disinfection methods are to be used.

Several methods can be used to evaluate the effectiveness of antimicrobial agents. However, before we look at the specific methods of evaluating them, let us review some of the terminology commonly used to discuss these agents and how they inhibit or destroy microorganisms.

BACKGROUND

Antimicrobial agents include all those agents, physical or chemical, that destroy or inhibit microorganisms. We examined the effectiveness of several physical control agents in Exercises 23 and 24. In this exercise, as well as in Exercises 26 and 27, we will concentrate on the effect of chemical control agents on microbial growth.

The term **germicide** is often used to refer to a chemical capable of destroying microorganisms. Increased specificity of action can be shown by the use of more precise terms, such as **bactericide** (an agent that kills bacterial cells, but not necessarily endospores), **sporicide** (an agent that kills bacterial endospores or fungal spores), **virucide** (an agent that kills viruses), or **fungicide** (an agent that kills fungi). The distinction made by these terms is important in determining their possible application. For example, what type of agent would you want to use to kill the mildew in your shower?

Other chemicals do not kill microorganisms, but simply inhibit their growth. These agents are referred to as being **microbistatic**. Again, it is possible to determine greater specificity of action by the utilization of the term **bacteriostatic** to describe those agents that specifically inhibit the growth of bacterial cells.

Chemical control agents are also classified by their ability to be used on living tissue in contrast to only being applicable to inanimate surfaces. Those chemicals that are applied to living tissue for the prevention of infections (antisepsis) are commonly referred to as being **antiseptics**. They must be relatively nontoxic, allowing their safe application to skin or mucous membranes, and they are usually, but not always, bacteriostatic. **Disinfectants** are used on inanimate surfaces. These may be the same agents as antiseptics, but they are usually used in a higher concentration. Would household bleach be considered an antiseptic or a disinfectant? How about hydrogen peroxide?

There is a third group of chemical control agents—the **chemotherapeutic agents**. We will be considering these in Exercise 27. These chemicals are used specifically to treat disease. They consist of antibiotics, or those chemicals derived from living agents, and synthetic drugs.

FACTORS AFFECTING ANTIMICROBIAL ACTIVITY

Several factors must be taken into consideration when choosing antimicrobial agents as they can affect the agent's effectiveness. What is the degree of microbial contamination? Are endospores likely to be present? How long will the organisms be exposed to the antimicrobial agent? What is the temperature of application? Does the microbial environment contain extraneous organic material? What is the pH of the microbial environment?

CLASSIFICATION OF ANTIMICROBIAL AGENTS

Antimicrobial agents are classified by their chemical composition and method of action. The major groups, their action and effectiveness are summarized in Table 25.1.

Usually when you receive an injection or have blood drawn, the skin is cleansed with an alcohol wipe. Because of the rapid evaporation rate, this has somewhat limited antimicrobial effectiveness, simply killing vegetative cells that are on the surface of the skin. Endospores, resistant bacteria, and organisms within the skin pores are not affected. Alcohol serves as a lipid solvent, dissolving cell membranes. When it is coupled with water it also denatures protein. For this reason, ethyl or isopropyl alcohol are most commonly used in a 70% to 80% concentration.

Another widely used group of skin

Table 25.1   Antimicrobial Agents and Their Applications

ANTIMICROBIAL GROUP	MODE OF ACTION	EFFECTIVENESS	EXAMPLES	APPLICATIONS
Alcohols	Denature proteins, dissolve lipids, dehydrate molecules	Kill vegetative cells but not endospores	Ethanol and isopropanol	Disinfect instruments and clean skin
Alkylating agents	Raise pH, denature protein	Kill vegetative cells and endospores	Formaldehyde Glutaraldehyde	Embalming and vaccines; Antiseptic (Cidex)
Dyes	May interfere with replication or block cell–wall synthesis	Bacteriostatic in high concen–trations; at lower concentrations only inhibits gram-positive organisms	Acridine Crystal violet	Clean wounds; treat some protozoan and fungal infections; selective agent in bacteriological media
Gaseous chemosterilants	Denature protein	Sterilizing agent, especially for materials which would be harmed by heat	Ethylene glycol Propolene glycol Ethylene oxide	Aerosols sterilize inanimate objects that would be harmed by high temperatures
Halogens	Inactivate protein, oxidize cell components	Kills vegetative cells and some endospores	Iodine, Chlorine	Antiseptic; surgical preparation (Betadine); disinfection for water, dairies, restaurants
Heavy metals	Denature protein	Disinfectant and antiseptic applications dependent upon concentration	Silver nitrate Mercury com–pounds Copper	Prevent gonococcal infections; disinfect skin and inanimate objects; inhibit algal growth
Organic acids	Metabolic inhibition, lower pH, denature protein	Widely used as fungicide	Sorbic acid Benzoic acid Calcium propionate	Food preservation, antifungal agents
Oxidizing agents (Peroxygens)	Oxidation, disrupt disulfide bonds	Especially effective against oxygen-sensitive anaerobes	Hydrogen peroxide Potassium perman–ganate Ozone	Clean puncture wounds; disinfect instruments; water disinfection
Phenolics	Denature protein, disrupt cell membranes	Kill vegetative cells but not endospores; many phenolics effective in presence of organics	Cresols Hexachlorophene Chlorhexidine gluconate Triclosan	Preservatives; antiseptic; surgical scrub; antibacterial soaps and detergents
Quatenary ammonium compounds	Denature protein, disrupt cell membranes	Kill most vegetative cells	Zephiran, Roccal, Cepacol, other cationic detergents	Sanitization in restaurants, laboratories, industry
Soaps and Detergents	Lower surface tension	Alone, primarily sanitizing agent, may be coupled with other antimicrobial agent(s)		Hand washing, laundering, sanitizing kitchen and dairy equipment

antiseptics as well as disinfectants are the halogens—primarily iodine and chlorine. Tincture of iodine was one of the first antiseptic agents used. It has been replaced in clinical applications by iodophors, which combine the iodine with organic surfactants. This provides a longer period of antisepsis. Betadine and Isodine are commonly used iodophors for surgical scrubs and preparation of skin prior to incisions. Chlorine is used as hypochlorous acid in bleach, water treatments, and various commercial sanitizing agents. One major limitation to its application is that it may be inactivated by the presence of organic material.

The heavy metals—mercury, copper, silver, zinc, and selenium—work by oligodynamic ("little power") action, combining with protein molecules and denaturing them. Frequently used forms of mercury have included merthiolate and mercurochrome for skin antisepsis, and trimerosal for skin antisepsis, instrument disinfection, and vaccine preservation. Mercury is rarely used today owing to its cumulative toxicity. Copper is most commonly used as copper sulfate for the control of algal growth in bodies of water. Silver nitrate has been used to prevent the growth of gonococcal organisms in the eyes of newborns, and silver nitrate and silver sulfadiazine are used for severe burns. Both zinc and selenium are included as antifungal agents in skin ointments (diaper rash and athlete's foot ointments) and shampoos.

Joseph Lister first used phenol as an antimicrobial agent. Since then, phenol has been replaced by a number of phenol derivatives or phenolics due to the toxicity of phenol. These agents denature protein and disrupt the membranes of cells. Their effectiveness is not affected by the presence of organic material. Many of the phenolics, such as hexachlorophene and chlorhexidine gluconate, are combinations of phenolics and halogens, providing increased effectiveness. One of the most commonly used phenolics today is triclosan. It is used widely in soaps and detergents. Take a look around your home and you will most likely encounter several cleansing agents that contain triclosan.

Many of the commonly used laboratory disinfectants are quaternary ammonium compounds. Included are Zephiran and Roccal. Their use is now somewhat limited, however, as they have been shown to not only fail to inhibit *Pseudomonas* organisms, but to support their growth.

METHODS OF EVALUATION OF ANTIMICROBIAL AGENTS

There are three major ways of evaluating the effectiveness of antimicrobial agents. Each of the techniques has certain advantages and certain disadvantages. All are useful under certain circumstances. These methods are summarized below.

Disk-Diffusion Technique

The disk-diffusion technique is commonly used to test the overall effectiveness of antiseptics, disinfectants and chemotherapeutic agents. Its application for chemotherapeutic agents will be used in Exercise 27 (the Kirby-Bauer procedure). The test is based upon the diffusion of molecules (the agent being tested) from the point of high concentration (the disk) to low concentration (the media). The effectiveness of the agent diminishes as the concentration decreases, resulting in varied zones of inhibition (see Figure 25.1). The rate of molecular diffusion is dependent upon several factors, including the molecular weight and shape of the substance being tested, its concentration, the media being used, the presence of organic material, and the temperature of incubation. For this reason one cannot make the assumption that the larger the zone of inhibition, the more effective an agent is, but only whether or not the chemical agent will inhibit the given organism. In addition, it is not possible to determine whether the agent is bactericidal (killing the microorganism) or simply bacteriostatic (inhibiting the organisms' growth). The disk-diffusion test is the easiest procedure to use when one is simply trying to determine the overall effectiveness of a given agent against a given organism.

Use-Dilution Technique

In the use-dilution procedure bacteria are removed from the inhibitory agent, enabling the determination of its bactericidal or bacteriostatic action. Varied dilutions of the antimicrobial agent can be used to determine the **minimal inhibitory concentration (MIC)** or **minimal bactericidal concentration (MBC)** of the agent. Because this technique is more complex than the disk-diffusion method, it is not routinely performed. It would be used for the establishment of effectiveness for a newly formulated antimicrobial, or clinically when patients do not respond to what is normally considered adequate therapy or who have a relapse while undergoing adequate therapy. We will utilize a modified use-dilution test in Exercise 26.

Phenol Coefficient Test

The phenol coefficient test determines the ratio of the concentration of antimicrobial agent being tested to the concentration of phenol that will kill the bacteria being tested in 10 minutes, but not in 5 minutes. This provides a quantitative comparison of the tested

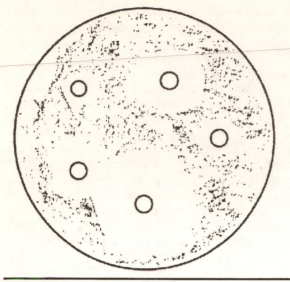

Figure 25.1   Zones of inhibition.

agent and phenol; however the value of this information is limited. It must be remembered that the phenol coefficient value determined is significant only for the organism tested under the test conditions used and should not be interpreted more broadly. It also does not take into account the presence of any extraneous organic material.

CHOOSING AN IDEAL DISINFECTANT
As we have seen, there are many different types of disinfectant or antiseptic agents. When selecting one for use several criteria should be considered. An ideal disinfectant should be:

1. highly effective against many organisms in dilute concentrations;
2. fast-acting even in presence of organic material;
3. non-toxic if applied topically, inhaled, or ingested;
4. stable;
5. soluble in water or alcohol;
6. able to penetrate material to be disinfected without damaging it;
7. colorless, odorless;
8. biodegradable; and
9. inexpensive and readily available.

No agent is likely to satisfy all of these criteria for the desired application. We must then choose the one agent that best satisfies our needs.

LABORATORY OBJECTIVES
In this exercise you will use the disk-diffusion technique to evaluate the effectiveness of several disinfectants and antiseptics. You should:

- distinguish between disinfectants and anti-septics.;
- distinguish between bactericidal and bacterio-static action of chemical agents;
- evaluate the effects of antiseptics or disin-fectants on selected bacterial cultures;
- discuss the limitations of using the disk-dilution method for evaluating chemical agents.

MATERIALS NEEDED FOR THIS LAB
1. Twenty-four hour broth cultures of:
    Escherichia coli
    Staphylococcus epidermidis
    Bacillus subtilis.
2. Media per group:
   Mueller-Hinton agar plates (3)
   Blood agar plates (3).
3. Varied antiseptics and/or disinfectants. At least six should be available for your use. You may also bring in any antimicrobial agent(s) you would like to test.
4. Petri dish containing sterile filter paper disks.
5. Sterile swabs.
6. Forceps.

LABORATORY PROCEDURE
1. Obtain one plate of Mueller-Hinton agar and one plate of blood agar for each organism being tested. Label with the test organism and your identifying information.
2. Dip a sterile swab into a broth culture of the test organism. Swab the entire surface of the Mueller-Hinton and Blood agar plates.
3. Repeat steps 1 and 2 for each organism being used.
4. Determine which antiseptics or disinfectants you will use. The same six agents should be used on each of your plates. Indicate on the bottom of the plate the position for each of the disks and the chemical agent present on the disk.
5. Using your sterile forceps, pick up one of the sterile filter-paper disks. Dip the edge of the disk into the desired solution and allow it to absorb the solution by capillary action.
6. Place the disk in the designated area on the inoculated petri plate. Be sure to gently press down the disc on the agar surface using your forceps. This will assure better adherence to the surface when the plates are inverted for incubation.
7. Repeat steps 5 and 6 for each of the test solutions on each of your plates, being sure to re-sterilize your forceps before each application.
8. Incubate your plates at 37°C until the next laboratory session.

9. Following incubation, the zones of inhibition should be determined for each solution on each plate. To determine the zones, measure either the diameter of the zone, or double the radius of the zone as measured in millimeters. Record your results on the Laboratory Report Form.

NOTES_____

Name _____

Section _____

LABORATORY REPORT FORM

EXERCISE 25
ASSAY OF ANTIMICROBIAL AGENTS: DISK-DIFFUSION METHODS

What is the purpose of this exercise?

1. For each of the disinfectants or antiseptics tested, determine the active ingredient. What class of inhibitory agent does it belong to and what is its mode of action?

CHEMICAL AGENT	ACTIVE INGREDIENT(S)	CLASS OF AGENT	MODE OF ACTION

2. Zones of inhibition (in mm):

CHEMICAL AGENT	Escherichia coli		Staphylococcus epidermidis		Bacillus subtilis	
	Mueller-Hinton agar	Blood agar	Mueller-Hinton agar	Blood Agar	Mueller-Hinton agar	Blood agar

3. Why was blood agar used in addition to Mueller-Hinton agar?

4. What variations in results did you observe when comparing effectiveness of the tested solutions for:

    a. the blood agar plate and the Mueller-Hinton plate for the same test organism?

    b. the varied organisms on the Mueller-Hinton plates?

QUESTIONS

1. What factors, other than the agent being tested, can influence the size of the zone of inhibition?

2. Were any colonies observed within the zone of inhibition? Explain why such colonies might be observed.

ASSAY OF ANTIMICROBIAL AGENTS: USE-DILUTION METHOD

We saw in Exercise 25 that the overall effectiveness of antimicrobial agents can be determined by using a disk-diffusion method. What could not, however, be determined by this method was whether the microorganisms were killed or simply inhibited. In this exercise we will determine the bactericidal or bacteriostatic nature of select antimicrobial agents.

BACKGROUND

If antimicrobial agents are bactericidal, they will irreversibly damage the cell so that the cell is no longer viable, whether the chemical agent is present or not. This reaction cannot be reversed. Bacterio–static agents, in contrast, are only effective when they are in direct contact with the microorganism. When the chemical agent is removed, the organism is again able to grow. In other words, the effect is reversible.

To prove bactericidal activity, the agent must be removed from contact by either washing the cells in saline or by diluting the agent down to a concentration that is no longer effective. Cells irreversibly damaged by the antimicrobial agent will not be able to grow when inoculated into a suitable growth medium in the absence of the agent.

STANDARD USE-DILUTION TEST

The standard method of testing for use-dilution is the American Official Analytical Chemist's use-dilution test. In this test, chemicals are tested against three different bacterial species—*Salmonella cholerasuis, Staphylococcus aureus,* and *Pseudomonas aeru–ginosa.* Metal rings are dipped into standardized cultures, removed and dried. They are then placed into the recommended concentrations (the concentration of use) of the disinfecting agent for 10 minutes at room temperature, after which they are transferred to appropriate growth media. Following incubation, they are examined for signs of microbial growth. Any growth would indicate bacteriostasis, and the absence of growth would indicate bactericidal activity.

LIMITATIONS TO THE
STANDARD TEST PROCEDURE

The standard test as outlined above can effectively demonstrate the bactericidal or bacteriostatic effect of a chemical agent, but it does not show how other factors can influence the antimicrobial effect of these chemicals. We saw in Exercise 25 that the presence of organic material can influence the effectiveness of an antimicrobial agent. Other important factors would include the penetration capabilities of the chemical agent and the time of exposure.

We will be performing a modification of the standard use-dilution test. Toothpicks and stainless steel pins will be soaked in a bacterial suspension and dried. They will then be placed for varied periods of time into one of the antimicrobial agents used in Exercise 25. The stainless steel pins are comparable to the disinfection of a non-porous surface, whereas the toothpicks will represent a porous surface.

Following exposure, the pins and toothpicks will be transferred to tubes of nutrient broth and incubated so that any viable bacteria will grow. The volume of the broth in the tubes (approximately 10 ml) is so large relative to the amount of chemical agent adhering to the surface of the pin or toothpick that they will be diluted to a non-effective concentration. If no growth is observed following incubation, the bacteria can be assumed to have been killed and the agent said to be bactericidal. If growth occurs in the tubes, the agent is bacteriostatic at best. Is it possible to have a zone of inhibition with the disk-diffusion test and growth with the use-dilution test? What about no zone of inhibition with the disk-diffusion test and growth in the use-dilution test? What could you conclude if there was both a zone of inhibition and no growth? No zone inhibition and growth?

LABORATORY OBJECTIVES

In this exercise you will be performing a modified use-dilution test using an antimicrobial agent that was also tested by the disk-diffusion method. Following the completion of this exercise you should:

- understand the difference between bacteriostatic and bactericidal agents;
- explain how to test to determine whether chemical are bacteriostatic or bactericidal;
- describe the use-dilution test;
- discuss those factors that could influence the results of a use-dilution test.

MATERIALS NEEDED FOR THIS LAB
1. Broth cultures of the following:
   Echerichia coli (24 hours)
   Micrococcus luteus (24 hours)
   Bacillus subtilis (72 hours)
2. One of the antimicrobial agents used in Exercise 25.
3. Two empty sterile petri plates.
4. Six toothpicks and six stainless steel pins.
5. Twelve tubes of sterile nutrient broth (approximately 10 ml per tube).
6. Filter paper.
7. Forceps.

LABORATORY PROCEDURES

NOTE: Each group will be assigned to use one of the microorganisms for the test protocol.

Figure 24.1 provides a flow chart for this procedure. Be sure to examine it carefully before starting.

1. Divide the nutrient broth tubes into two sets— one for use with the toothpicks and one with the stainless steel pins.
2. Label the nutrient broth tubes as follows:
   #1—Toothpick, 0 Time (control)
   #2—Toothpick, 1 minute soak
   #3—Toothpick, 2 minute soak
   #4—Toothpick, 5 minute soak
   #5—Toothpick, 10 minute soak
   #6—Toothpick, 15 minute soak
   #7—Pin, 0 time (control)
   #8—Pin, 1 minute soak
   #9—Pin, 2 minute soak
   #10— Pin, 5 minute soak
   #11—Pin, 10 minute soak
   #12—Pin, 15 minute soak

3. Pour 10 ml of the assigned bacterial culture into one of the petri dishes. Add the toothpicks and stainless steel pins and allow them to soak for about 5 minutes. Be sure that they are covered by the broth suspension.

NOTE: When making the transfers into your nutrient broth tubes, drop the toothpick or pin into the broth. Do not allow the forceps to touch the broth in the tubes. Be sure to sterilize your forceps between each use.

4. Using your forceps, remove the toothpicks and pins from the broth and place them on a piece of filter paper to eliminate any excess of broth culture.
5. Transfer one toothpick into tube 1 and one pin into tube 7. These are your "zero time" or control tubes. What is the purpose of using these controls?
6. Pour 10 ml of the antimicrobial agent into the second petri dish. Add the remaining five toothpicks and five pins to the dish. Start timing as soon as they are added.
7. Transfer one toothpick and one pin to the appropriate nutrient broth tubes at the designated time intervals. Remember the time is determined from the minute they are placed into the antimicrobial agent.
8. Place all tubes in the incubator.
9. Be sure to carefully dispose of the petri plates as indicated by your instructor.
10. Following incubation, record whether or not growth occurred in each tube. Compare the results for your organism with those for the other organisms. Record the results on your Laboratory Report Form.

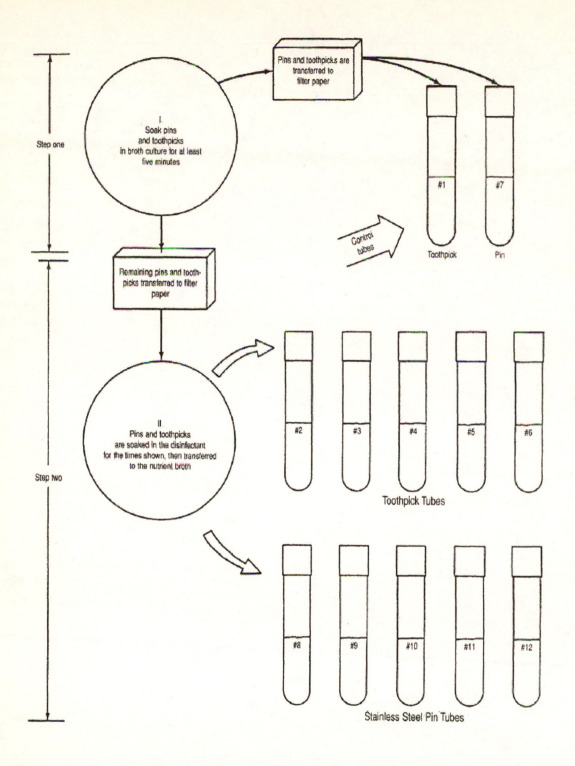

Figure 24.1   Use-dilution assay procedure.

NOTES

LABORATORY REPORT FORM

EXERCISE 26
ASSAY OF ANTIMICROBIAL AGENTS: USE-DILUTION METHOD

What is the purpose of this exercise?

1. Antimicrobial agent tested: _____

2. Organism tested against: _____

3. What are the active ingredient(s) in the antimicrobial tested?

4. Complete the following table. Indicate any growth as (+).

TIME	Escherichia coli		Micrococcus luteus		Bacillus subtilis	
	Toothpick	Pin	Toothpick	Pin	Toothpick	Pin
0						
1 min						
2 min						
5 min						
10 min						
15 min						

5. Would you consider the antimicrobial agent tested to be bacteriostatic or bactericidal?

6. Could you determine from these results if the agent is sporicidal? Explain why or why not.

QUESTIONS

1. Explain any differences that might be observed between the killing times noted for the pins and toothpicks.

2. What test(s), besides the use-dilution test, would you need to employ to determine the effectiveness of an antiseptic agent?

3. What factors would contribute to varied results for *Micrococcus luteus* and *Escherichia coli*? For *Bacillus subtilis* and *Micrococcus luteus*?

ANTIBIOTIC SENSITIVITY TESTING: THE KIRBY-BAUER PROCEDURE

Successful isolation and identification of clinically significant bacteria from a clinical specimen is only half the task expected of the microbiology laboratory. Virtually every isolate that may have clinical significance must be tested to determine its antibiotic sensitivity pattern.

There are two reasons for this: First, the physician needs to have the sensitivity pattern available to protect the patient against the possibility that this particular isolate may be resistant to the commonly used antibiotics. Second, antibiotic sensitivity patterns have proven to be consistent for a given organism and can be used as an additional diagnostic characteristic.

BACKGROUND

Alexander Fleming first noted in 1929 that *Penicillium* growing on an agar plate could inhibit the growth of *Staphylococcus aureus*. This was followed by Selman Waksman discovering that *Streptomyces griseus* produced streptomycin, which inhibited many gram-negative organisms. The additional discovery in the 1930s that sulfanilamide would also inhibit certain infectious agents began the era of chemotherapy and our ability to cure microbial infections.

Today we have a large number of antimicrobial agents available for treating infections. These have become an essential part of modern medical practice by using selective toxicity to eliminate infecting organisms or preventing the establishment of infection.

Antibiotics, derived from the term *antibios* (against life), were originally defined as those chemicals produced by microorganisms that inhibit the growth of other microorganisms. This included only those naturally occurring agents, making a distinction between them and the synthetic chemical agents, such as the sulfonamides. Today we find that the distinction between the natural and synthetic drugs has pretty much disappeared as synthetic versions of naturally occurring antimicrobials are being manufactured. Most of these drugs used today are referred to as antibiotics.

Antibiotics that are effective against a wide range of organisms, including both gram-positive and gram-negative species, are referred to as being broad-spectrum. **Narrow-spectrum** drugs are effective against a very select group of organisms. In all cases, the use of antibiotics is dependent upon the exhibition of **selective toxicity**. This is achieved by involving biochemical features of the pathogen that are not possessed by the host cells, making the antibiotic toxic to the microorganism but not to the host. The five major modes of action utilized by antimicrobial drugs are (1) the inhibition of cell–wall synthesis, (2) the disruption of cell membrane functioning, (3) the inhibition of protein synthesis, (4) the inhibition of nucleic acid synthesis, and (5) competitive inhibition of metabolic pathways. See your textbook for further discussion of these modes of action.

An emerging problem today is the rapid development of microorganisms resistant to the effects of chemotherapeutic agents. The following factors have been cited for contributing to this problem:

- development of alternative biochemical pathways by microorganisms, circumventing the effect of the chemotherapeutic agent;
- production of enzymes by microorganisms that inactivate the chemotherapeutic agent;
- presence of drug-resistant plasmids that provide resistance to multiple drugs for some bacteria and that can be spread from one organism to another;
- undertreatment of infection, resulting in the survival of the most resistant strains;
- overuse of antibiotics, resulting in the increased prevalence of resistant organisms through the widespread destruction of susceptible strains.

PPNG (penicillinase-producing *Neisseria gonorrhoea)* and MRSA (methicillin-resistant *Staphylococcus aureus*) are just two of the resistant forms being frequently encountered today. We will continue to see more species becoming resistant and the re-emergence of pathogens previously considered "under control" (i.e., *Mycobacterium tuberculosis* and vancomycin-resistant *Staphylococcus aureus* or VRSA) if both the overuse and the unnecessary uses of antibiotics continue.

Today the need for careful testing of pathogens for drug sensitivity is becoming more essential. Two major tests are used to determine the susceptibility of organisms to chemotherapeutic agents—the Minimum Inhibitory Concentration (MIC) test and the Kirby-Bauer disk-diffusion test. The Kirby-Bauer test is the most common test utilized and the procedure that we will examine in this exercise. See your textbook for further information about the MIC test.

THE KIRBY-BAUER DISK DIFFUSION TEST

Diffusion tests have been used for some time to determine the effect of various chemical agents on bacterial growth. Originally, wells were cut in agar plates and the chemical agent was inserted into the well. This was replaced by the use of paper disks containing the chemical agent following their introduction in 1947 by Bondi. Disk-diffusion tests, as they became know, had little credibility, however, until Bauer, Kirby, Sherris, and Turck developed correlations between MIC results and those achieved by a standardized disk–susceptibility test. By correlating the observed zones of inhibition with the known MIC values, they were able to standardize zone interpretations as indicating **susceptibility** (S) or **resistance** (R) by the given organism. If the zone size did not correlate well with the MIC standards, the results were considered **inconclusive** (I). The Kirby-Bauer test has become the standard method for testing antibiotic susceptibility in clinical laboratories. Table 27.1 shows the correlation between zone of inhibition and susceptibility for some of the antibiotics commonly used.

The controls used in the Kirby-Bauer test include:

- **Standardized bacterial suspension.** The amount of antibiotic needed to inhibit a population of bacteria is proportional to the size of the population. The density of the bacterial suspension used in the Kirby-Bauer procedure is maintained constant by comparing it to a standardized suspension of a known turbidity. The standard used is the McFarland Standard Tube No. 0.5.
- **Standardized concentration of antibiotics in the disks.** All other things being the same, the diffusion rate of any chemical is directly proportional to the concentration of that chemical. The amount of antibiotic in the disk must be standardized if any consistency in the zone of inhibition is to be expected. Stan–dardized disks are available from biological supply houses.
- **Standardized medium.** The Kirby-Bauer procedure uses Mueller-Hinton agar. This medium was chosen because its properties relative to diffusion of antibiotics and growth of bacteria are known. Any other medium would not necessarily allow the antibiotic to diffuse at the same rate and would not support the same amount of bacterial growth. The depth of the medium must also be standardized. (Why would the depth of the medium affect the size of the zone of inhibition observed?)
- **Standardized incubation time and temperature.** The rate of diffusion is temperature dependent and increases as the temperature is increased. Furthermore, if the time for diffusion is extended indefinitely, the concentration of the antibiotic in the medium will eventually be equal everywhere. Antibiotics are bacteriostatic. If the concentration of the antibiotic falls below the MIC value, the organism will resume growth. An incubation temperature of 35° to 37°C for no more than 24 hours (usually 16 to 18 hours) is required.

Following incubation, the diameter of the zone of inhibition is measured and this value is compared with the standard table to determine the results as susceptible, resistant, or inconclusive for the given antibiotic and organism.

- **Susceptible.** The minimum zone diameter, expressed in millimeters, that indicates the organism will be sensitive to the antibiotic at normal therapeutic doses. If a zone diameter larger than this value is observed, the organism is considered to be sensitive to the antibiotic.
- **Resistant.** The minimum zone diameter, expressed in millimeters, that indicates the organism will be resistant to the antibiotic at normal therapeutic doses. If a zone smaller than this value is observed, the organism is considered to be resistant to the antibiotic.
- **Inconclusive or intermediate.** The range of zone diameters, expressed in millimeters, between the sensitive and resistant zone diameters where individual differences between patients and bacterial strains may cause one particular combination to be resistant while another combination might prove to be sensitive.

Examine Table 27.1. Note that for some of the antibiotics (ampicillin and penicillin G) the proper evaluation of the zones is dependent upon the organism being tested. Why do you think there are higher values required for some organisms than for others?

If any growth of colonies occurs within the zone of inhibition, they must be gram stained to determine if they are contaminants or a mutant (resistant) strain of the pathogen.

The data determined by the Kirby-Bauer test provides the physician with a screening test to determine which drugs are exhibiting significant antibiotic activity. Final decision on the antibiotic of choice is not, however, necessarily going to be that drug that produces the largest zone of inhibition. Such additional factors as potential of side effects, biotransformation potential (whether the anti–microbial agent will remain active in the body long enough to be selectively toxic to the pathogen), and the ability of the antimicrobial agent to reach the site of the infection in adequate concentration must all be taken into account.

Kirby-Bauer sensitivity tests are not done routinely on all infectious agents. Cost and time are two major reasons they are not performed. It has been shown that if all infections are cultured and a sensitivity test run on them guaranteeing that antibiotics are only being used when necessary, there can be a significant decrease in the development of antibiotic–resistant bacterial cultures. This, plus the education of the overall population about the proper use of antibiotics, will enable us to continue using antibiotics for the treatment of bacterial infections.

LABORATORY OBJECTIVES

In this exercise, you will be determining the antibiotic-sensitivity pattern for several organisms. In performing this test you will

- perform the Kirby-Bauer test for the determination of antibiotic sensitivity/ resistance;
- distinguish between broad-spectrum and narrow-spectrum antibiotics;
- determine the presence of drug-resistant organisms;
- interpret the results of the antibiotic sensitivity test based on the zone size;
- discuss the medical importance of the antibiotic-sensitivity test.

Table 27.1   Interpretive Standards for Disk-Diffusion Susceptibility Testing

ANTIMICROBIAL AGENT	DISK CONTENT	ZONE OF INHIBITION (mm)		
		Resistant	Intermediate	Susceptible
Ampicillin				
Enterobacteriaceae	10 μg	≤ 13	14–16	≥ 17
Staphylococci		≤ 28		≥ 29
Enterococci		≤ 16		≥ 17
Streptococci		≤ 21	22–29	≥ 30
Bacitracin	10 units	≤ 8	9–12	≥ 13
Cephalothin	30 μg	≤ 14	15–17	≥ 18
Chloramphenicol	30 μg	≤ 12	13–17	≥ 18
Clindamycin	2 μg	≤ 14	15–20	≥ 21
Erythromycin	15 μg	≤ 13	14–22	≥ 23
Gentamicin	10 μg	≤ 12	13–14	≥ 15
Kanamycin	30 μg	≤13	14–17	≥ 18
Methicillin	5 μg	≤ 9	10–13	≥ 14
Neomycin	30 μg	≤ 12	13–16	≥ 17
Nitrojurantoin	300 μg	≤ 14	15–16	≥ 17
Penicillin G				
Staphylococci	10 units	≤ 28		≥ 29
Enterococci		≤ 14		≥ 15
Streptococci		≤ 19	20–27	≥ 28
Polymyxin B	300 units	≤ 8	9–11	≥ 12
Rifampin	5 μg	≤ 16	17–19	≥ 20
Streptomycin	10 μg	≤ 12	12–14	≥ 15
Sulfisoxazole (Gantrisin)	250 μg	≤ 12	12–16	≥ 17
Sulfonamides	300 μg	≤ 12	13–16	≥ 17
Tetracycline	30 μg	≤ 14	15–18	≥ 19
Vancomycin				
Enterococci	30 μg	≤ 14	15–16	≥ 17
Other gram (+)		≤ 9	10–11	≥ 12

MATERIALS NEEDED FOR THIS LAB
1. Broth cultures of:
   Staphylococcus aureus
   Escherichia coli
   Pseudomonas aeruginosa.
2. Three Mueller-Hinton agar plates poured to a depth of 4.0 mm.
3. Sterile swabs.
4. Antibiotic disks for eight different antibiotics in either single disk or multiple disk dispenser (see Figure 27.1).
5. Forceps.

LABORATORY PROCEDURE
1. Obtain and label one Mueller Hinton plate for each organism being tested.
2. Note the code present on the antibiotic disks. You do not need to label your plates for the location of the disks.
3. Swab the surface of each plate with the organism being tested. Be careful to cover the entire surface of the plate so that you will get a solid, continuous lawn of bacteria. Dispose of your swab as indicated by your instructor.

4. Place the antibiotic disks on the plates.
   a. If a multiple dispenser is used, place it over the plate and slowly depress the plunger or slide the lever to release the disks. Using your sterile forceps, press lightly on each disk to be sure that it will adhere to the surface of the medium when the plate is inverted. Repeat for each plate.
   b. If using single–disk dispensers, carefully apply each of the disks to your plates, being sure to keep them equally spaced. Using your sterile forceps, press lightly on each disk to be sure that it will adhere to the surface of the medium when the plate is inverted. Repeat for each plate.
5. Incubate the plates for 24 hours. If you are unable to read the results at that time, the plates should be refrigerated until the next laboratory period.
6. Measure (in mm) the zone of inhibition for each antibiotic.
7. Record your results on the Laboratory Report Form. Be sure to note the presence of any colonies within the zone of inhibition.

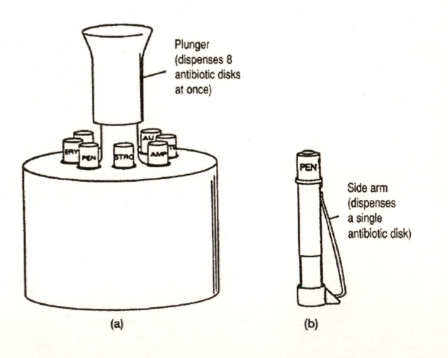

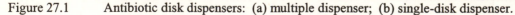

Figure 27.1   Antibiotic disk dispensers: (a) multiple dispenser; (b) single-disk dispenser.

LABORATORY REPORT FORM

EXERCISE 27
ANTIBIOTIC SENSITIVITY TESTING: THE KIRBY-BAUER PROCEDURE

What is the purpose of this exercise?

1. Complete the following chart:

CHEMOTHERAPEUTIC AGENT	DISK CODE	*Staphylococcus aureus*		*Escherichia coli*		*Pseudomonas aeruginosa*	
		Zone (mm)	S, I, or R	Zone (mm)	S, I, or R	Zone (mm)	S, I, or R

2. Based on your results, which chemotherapeutic agent(s) tested were broad–spectrum? Narrow–spectrum?

3. Were any colonies growing within any of your zones of inhibition? If so, how would you determine if it was a contaminant or a resistant colony?

QUESTIONS

1. What do the following terms mean?
   a. PPNG –

   b. MRSA –

2. How does the MIC test differ from the Kirby-Bauer test?

3. Were all of the conditions of a standardized Kirby-Bauer test met as you performed this assay? If not, which were not?

4. What is the significance of colonies that develop within otherwise clear zones of inhibition? If the laboratory report for one of your patients indicated colonies within the zone, what concerns would you have for your patient?

5. Drawing upon the results of this exercise, why is *Pseudomonas aeruginosa* of such concern in burn patients and in immunologically compromised patients?

DETECTION OF MUTANT STRAINS OF BACTERIA

In this exercise we are going to explore several methods of identifying the presence of genetic variations in bacteria. These variations may be either induced (due to exposure to a known mutagenic agent) or spontaneous (naturally occurring without any known exposure to a mutagenic agent).

BACKGROUND

Mutations are changes in the base sequence of DNA that are transmitted to daughter cells and which may result in an inheritable change in the phenotype of a cell. Spontaneous mutations occur at a relatively low rate, varying from 1 in 10^4 to 1 in 10^{12} cell divisions. When we consider the rapid rate of cell division that occurs in bacteria, we can see that there are likely to be a number of mutants found in any bacterial population. For example, one gene present in *Escherichia coli* is the Gal+ gene. This gene enables the organism to metabolize galactose. It will undergo a spontaneous mutation to the Gal− form, (unable to utilize galactose), at a rate of approximately 2 in 10^7 cell divisions.

If the new phenotype (visible trait) is selected against by the environment (natural selection), the cell will disappear from the population. If, however, the change is beneficial to the organism's survival, the new cell type may gradually increase in number and eventually may become a significant part of the population. A neutral mutation—one that is neither selected for or against—will have little, if any, effect on the distribution of the mutant in the population.

The development of a mutant population, then, is dependent upon two processes: the mutation itself and the selection of that mutant by the environment. An example of this would be the development of drug resistance in *Staphylococcus aureus*. The ability to synthesize penicillinase, an enzyme that inactivates most penicillins, was of no particular advantage to *Staphylococcus aureus* before penicillin became a widely used antibiotic. Penicillin did not become the well-known "wonder drug" until World War II when it was used extensively to treat wounded soldiers. After the war it became readily available to the public and was widely (indiscriminately?) used to treat virtually any bacterial infection and even viral infections such as a bad cold.

No one knows when the mutation for penicillin resistance first appeared in the population of *Staphylococcus aureus*. Extensive selection for penicillinase-positive bacteria could not occur as long as penicillin was not widely distributed. As the antibiotic came into greater usage, the mutant (drug-resistant) strain was suddenly placed at an advantage. It could inactivate the antibiotic, while the non-mutant (wild type) cells could not. Selection for the resistant members of the population was so extensive that it is now relatively uncommon to encounter an isolate of *Staphylococcus aureus* that is sensitive to the original penicillin G. Because of the continued overuse of antibiotics, we are currently seeing a significant amount of drug resistance in *Pseudomonas aeruginosa* and *Mycobacterium tuberculosis* infections in addition to that discussed in *Staphylococcus aureus*.

If a mutation results in the inability of an organism to synthesize a needed enzyme, it will have a nonfunctional metabolic pathway. If the pathway affected is for the production of a substance essential for the growth of the cell, the cell will die unless the needed substance is provided in its medium. These are considered nutritional mutants.

Mutant strains cannot be detected unless we expose the bacteria to an environment that would select for that mutant. How would you detect penicillin-resistant bacteria unless you exposed the population to penicillin? While this seems self-evident, it does raise an interesting question: How can you prove that the mutation occurs independently from its selection?

The selection of mutant cells can be by *direct selection*, where only those cells exhibiting the mutation can grow, or by *indirect selection*. An example of direct selection would be the addition of an antibiotic to the growth media for an organism that is normally inhibited by the antibiotic. Any colonies that grow would be due to the presence of the mutation. In indirect selection, bacteria are grown in an environment that is not selective for a given mutation. A replica plate is used to transfer members of each colony to a selective environment (see Figure 28.3). By comparing the presence of colonies following incubation we can indirectly determine the mutated colonies by their presence or absence in the selective environment.

In this exercise, we will utilize both direct and indirect selection of mutated colonies. We will also be detecting the presence of spontaneous mutations and the existence of induced mutations.

LABORATORY OBJECTIVES

Bacterial mutations appear randomly in the population at a predictable rate. The mutation rate can be increased with certain mutagenic agents, but the resulting mutations are still randomly determined. A mutant population must be selected for by applying the correct environmental conditions. You should:

- understand the relationship between *mutation* and *selection*;
- understand that mutation rates can be altered by exposure to mutagenic agents, but that the randomness of these mutations cannot be altered;
- understand the relationship between the indiscriminate use of antibiotics and the development of resistant strains of bacteria;
- distinguish between direct and indirect selection of mutations;
- be able to use the replica plating technique.

MATERIALS NEEDED FOR THIS LAB

A. SPONTANEOUS MUTATION DETECTION

1. Nutrient broth suspensions of one of the following:
   Serratia marcescens
   Escherichia coli
2. Four nutrient agar talls
3. 50°C water bath
4. Antibiotic stock solutions. The stock solution should be prepared so that when 1 ml is added to a nutrient agar tall it will exceed the minimal inhibitory concentration (MIC) for the organism you are testing. You will use streptomycin and two other antibiotics.

Antibiotic	Stock Solution Concentration	Medium Concentration*
Penicillin	1000 mcg/ml	50.0 mcg/ml
Streptomycin	1000 mcg/ml	50.0 mcg/ml
Ampicillin	500 mcg/ml	25.0 mcg/ml
Polymyxin B	500 mcg/ml	25.0 mcg/ml
Erythromycin	1000 mcg/ml	50.0 mcg/ml
Gentamycin	100 mcg/ml	5.0 mcg/ml
Tetracycline	250 mcg/ml	12.5 mcg/ml
Chloramphenicol	500 mcg/ml	25.0 mcg/ml
Kanamycin	250 mcg/ml	12.5 mcg/ml

*These values are based on estimated in vivo dosage. Other values based on published MIC data

5. One tube of sterile water
6. Sterile 1.0 ml pipettes (5)
7. Sterile petri plates (4)
8. Bent glass spreader
9. 95% ethanol for sterilizing bent glass spreader
10. Beaker

B. INDUCED MUTATION DETECTION

1. Nutrient broth suspension of one of the following:
   Serratia marcescens
   Escherichia coli.
2. Four nutrient agar plates.
3. One minimal salts with glucose agar plate.
4. Three 99-ml sterile water dilution blanks.
5. Sterile 1.0-ml pipettes.
6. Bent glass spreader (Figure 28.1).
7. 95% ethanol for sterilizing bent glass spreader.
8. UV light.
9. Replica plate apparatus.
10. Sterile velveteen square.
11. Rubber band.

LABORATORY PROCEDURE

A. SPONTANEOUS MUTATION DETECTION

1. Obtain four sterile petri plates. Label three of them with the antibiotic, its concentration, and the name of the organism being tested. The fourth plate should be labeled "Control."
2. Melt tubes of nutrient agar and cool in a 50°C water bath.
3. Obtain one tube of melted agar. Aseptically add 1.0 ml of the streptomycin stock solution to the tube. Mix and pour the medium into the appropriately labeled petri plate. Allow it to solidify.
4. Repeat step 3 with two more antibiotic solutions.
5. The control is prepared by aseptically adding 1.0 ml of sterile water to a tube of medium, mixing and pouring it into the plate labeled "Control." Let it solidify.
6. Aseptically transfer 0.1 ml of the bacterial suspension to each of the agar plates.
7. Sterilize the bent glass spreader by dipping it in 95% ethanol, igniting the alcohol in the Bunsen burner flame, and burning off the alcohol. Let the spreader cool.

NOTE: When sterilizing the bent glass spreader, be sure that you do not hold the rod over anything flammable. Ignited alcohol can drop off the rod and set fire to whatever it lands on. Clear the countertop of all unnecessary materials. Always hold the flaming area downwards.

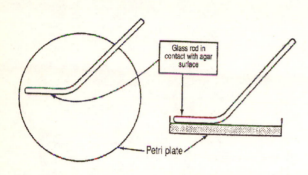

Figure 28.1   Use of bent glass spreader

8. Quickly spread the suspension over the surface with the sterilized bent glass.
9. Incubate the *E. coli* plates at 37°C and the *Serratia marcescens* plates at room temperature for 24 hours. If the results cannot be read at 24 hours, refrigerate the plates until the next laboratory period.
10. Count the colonies that develop on each of the plates. How can you determine that they represent antibiotic-resistant colonies of your test organism and are not contaminants?
11. Record your results on the Laboratory Report Form, Part A.

B. INDUCED MUTATION DETECTION

1. Obtain three nutrient agar plates. Label each with the organism used and the dilution (1:10,000; 1:100,000; or 1:1,000,000).
2. Label your sterile water blanks "1," "2," and "3."
3. Prepare a dilution series of your organism. Refer to Exercise 10, and to Figure 28.2.
    a. Aseptically transfer 1.0 ml of your broth suspension to the first 99 ml sterile water blank. Mix it well. (What is the dilution in this bottle?)
    b. Using another pipette, transfer 1.0 ml from water blank "1" to sterile water blank "2." Mix it well. (What dilution do you now have?)
    c. Using another pipette, transfer 1.0 ml from water blank "2" to sterile water blank "3." Using the same pipette, also transfer 1.0 ml from water blank "2" to the surface of the 1:10.000 plate and 0.2 ml to the surface of the 1:100,000 plate.
    d. Mix water blank "3" and transfer 1.0 ml to the surface of the 1:1,000,000 plate.

4. Sterilize the bent glass spreader by dipping it in 95% ethanol, igniting the alcohol in the Bunsen burner flame, and burning off the alcohol. Let the spreader cool.
5. Spread the dilution over the surface of the plate with the bent glass spreader. Repeat with the remaining plates.
6. Place each of your inoculated nutrient agar plates under the UV light for 30 seconds.
7. Invert the plates and incubate until the next laboratory period. The *Serratia marcescens* should be incubated at room temperature, the *E. coli* at 37°C. Determine the plate with 30 to 100 isolated colonies. This is your master plate that you will be using for the remainder of the exercise. Place a reference mark on the bottom of the plate (see Figure 28.3).
8. Obtain a plate of nutrient agar and of minimal salts with glucose agar. Label them and place a reference mark on the bottom of each.
9. Get a replica plating block. Put a piece of sterile velveteen cloth over the top. Touching only the corners of the cloth, bend it down over the surface of the block. Secure the cloth with a rubber band.
10. Inoculate your sterile agar plates in one of the following ways:
    a. Place the master plate and the uninoculated plates on the table with the reference marks all at 12 o'clock. Remove the covers. Carefully press the replica plate block to the surface of the master plate, aligning the handle with the reference mark. Without altering the position of the replica plate, touch it to the surface of the minimal salts with glucose agar plate and then to the nutrient agar plate. You should not go back to the master plate between inoculating your plates.
    b. Hold the replica plate device on the table with the velveteen surface up. Remove the lid from the master plate and invert it over the block, allowing the agar surface to lightly touch the velveteen surface. Replace the cover. Repeat with the nutrient agar plate.
11. Place the velveteen square in an autoclavable container as indicated by your instructor.
12. Incubate the plates as before. Refrigerate the master plate.
13. Following incubation, compare the presence of colonies on the master plate, nutrient agar plate, and minimal salts with glucose plate.

14.   Record your results on the Laboratory Report
       Form, Part B.

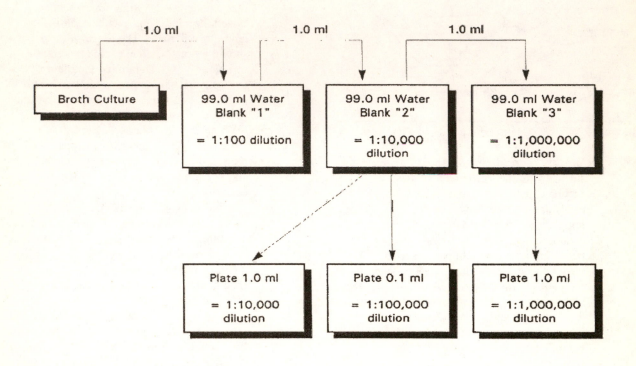

Figure 28.2   Preparation of dilution series

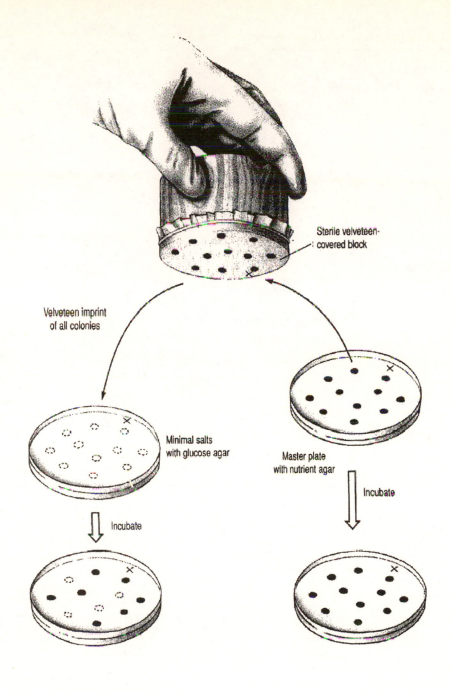

Sterile velveteen-covered block

Velveteen imprint of all colonies

Minimal salts with glucose agar

Master plate with nutrient agar

Incubate

Incubate

Figure 28.3   Use of replica plate apparatus

NOTES_____

LABORATORY REPORT FORM

EXERCISE 28
DETECTION OF MUTANT STRAINS OF BACTERIA

What is the purpose of this exercise?

A. SPONTANEOUS MUTATION DETECTION

Organism used: _____

Describe the growth present on the "Control" plate.

Complete the following table:

ANTIBIOTIC	CONCENTRATION IN MEDIUM	NUMBER OF RESISTANT COLONIES

How can you determine that the colonies seen are resistant mutants and not contaminants?

B. INDUCED MUTATION

Organism used: _____

Were any changes present in the appearance of the colonies following exposure to the UV light and initial incubation?

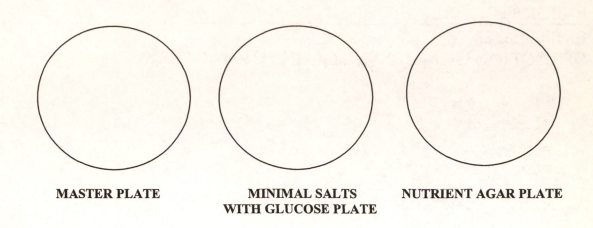

MASTER PLATE   **MINIMAL SALTS**   **NUTRIENT AGAR PLATE**
                       WITH GLUCOSE PLATE

Number of nutritional mutants observed: _____

QUESTIONS

1. What environmental factors would contribute to the selection of spontaneous mutations?

2. What environmental factors would contribute to the induction of mutations in organisms?

3. Why did we use both the nutrient agar and the minimal salts plus glucose medium for the replica plating?

4. Distinguish between direct and indirect selection of mutations.

EPIDEMIOLOGY

For a pathogen to cause disease, it must come in contact with a susceptible host. This requires a method of transmission and a suitable portal of entry for the organism to gain access to the host. The study of the occurrence, distribution and control of these infectious diseases, as well as non-infectious diseases, is the science of **epidemiology**. Involved is not only determining the causative agent of the disease, but how it was transmitted and how this spread can be slowed or stopped. In this exercise we will be simulating an epidemic with a known method of transmission.

BACKGROUND

Communicable diseases are characterized by their spread or transmission from person to person. The method of transmission may be by direct contact (sexually transmitted diseases), aerosol (influenza virus), water or food (cholera or *Salmonella*), fomites (inanimate objects, such as a dirty nail, which can transfer a pathogen to a host), or vectors (a living organism, such as a fly or mosquito, which can transmit a pathogen to a host).

If the disease, such as pneumonia, is constantly present in relatively low numbers, it is considered to be **endemic**. Often endemic diseases have reservoirs, or a site where the infectious agent can live without causing disease, and from which infection of others can occur. These reservoirs may be other animals (mice for the hanta virus, rats for typhus, cows for tuberculosis), or asymptomatic humans (**carriers** such as Typhoid Mary).

At other times, the **incidence** or **morbidity** of a disease (the number of cases in a population) may suddenly increase. This would be considered an **outbreak** or, if the incidence becomes unusually high in a localized region, an **epidemic** of the given disease. An epidemic may be a **common-source epidemic** in which a large number of individuals become ill from a single contaminated source (i.e., food poisoning following a company picnic). This will usually result in a very sudden increase, and later decrease in the incidence of the infection. If the rise and subsequent fall in the number of infected individuals is slower, it is often a **person-to-person epidemic** (i.e., influenza).

It is the role of the **epidemiologist** to determine the source(s) of an epidemic, how it has

been spread, and how it can be controlled. Today's epidemiologist is not only interested in the incidence and spread of infectious disease, but also the cause of non-communicable diseases (cancer, heart disease, or the effect of environmental toxins) and injuries (extent of injuries when seat belts are used) in groups of people.

EPIDEMIOLOGICAL METHODS

To monitor the occurrence of disease, the epidemiologist relies heavily on statistical data. What is the **incidence rate** of the given disease? This involves determining the number of people who develop the disease in question during a certain period of time and dividing that value by the total number of people in the population. An increase in the incidence rate would indicate that the epidemic is also growing. It is also important to know the **prevalence rate** of the disease. The prevalence rate differs from the incidence rate in that it measures the exact number of people who have a given disease at a particular time. This value indicates the magnitude of the epidemic. The source of this data might be public health records (vital statistics and census data), hospital records, questionnaires, or surveys.

Epidemiological studies may be descriptive, surveillance, field, or hospital studies. In **descriptive studies**, the location, time, individuals, procedures, and any common features are described. The morbidity (incidence of disease) and mortality (incidence of death) rates assist in providing a perspective on the increase or decrease of the disease and why the numbers are changing. Descriptive studies tend to be retrospective—looking back over what has already occurred. Their findings are often able to be used to improve overall health care, taking into account not only the disease itself but also those environmental factors contributing to its increased incidence.

Surveillance epidemiology tracks epidemic diseases and examines what changes are necessary to control its spread. These factors can include environmental change (improved water treatment methods), isolation of individuals who have been exposed, or vaccination. Surveillance methods were responsible for the eradication of smallpox and the control of polio. Surveillance also enables the rapid detection of re-emerging diseases, such as the drug-resistant strain of malaria.

When unexpected outbreaks of a disease occur, it often necessitates **field epidemiology** methods. Recent field studies include the identification of *E. coli* 0154:H7 food poisoning and the hanta virus outbreak in the southwest United States. In both of these studies, it was the sudden increase in illness noted by hospitals that triggered the study. At other times, citizen complaints can be the initiation of a field epidemiologic study. An example is the increased incidence of arthritis noted around Old Lyme, Connecticut, which led to the identification of Lyme disease.

The final type of epidemiological study is the **hospital study**. **Nosocomia**l, or hospital-derived, infections are a major concern to health care facilities. The hospital environment is especially vulnerable to the appearance of infections for a number of reasons. Many patients have weakened resistance to infectious disease due to their illnesses. Patients, themselves, are often reservoirs of highly virulent pathogens. Cross-infection can readily occur because of the crowding of patients in rooms. Hospital personnel move between rooms and can serve as a vector of transmission. Many hospital procedures are invasive, overcoming the body's natural defense barriers. Often patients are receiving medication that is immunosuppressive, weakening their ability to fight off infections.

Finally, the hospital environment contains many antibiotic–resistant organisms due to the frequent use of antibiotics. The result is that approximately 5% of all patients who are hospitalized will develop a nosocomial infection. To deal with these infections effectively requires the rapid identification of cases and the **etiology** (causative) agent, the characterization of the epidemiologic features of the infection, the development and implementation of control methods, and constant monitoring to assure that control methods are being used effectively.

PUBLIC HEALTH AGENCIES

Overseeing the occurrence of illness and its control are a number of public health agencies and private health organizations. These include the World Health Organization, the U. S. Department of Health and Human Services, the U. S. Public Health Service (the parent organization of the Center for Disease Control and Prevention, or CDC), and state and local public health departments. The presence of national and international monitoring organizations facilitate the development of wide-range environmental and reservoir eradication control methods. In our fast-traveling global society, we can no longer think of

ourselves as isolated for those epidemics occurring on the other side of the world.

LABORATORY OBJECTIVES

In this exercise you will be simulating an epidemic among your laboratory group. Through this experience you should:

• determine the source of a simulated epidemic;

• understand how the science of epidemiology contributes to our understanding of disease;

• understand how epidemiologists collect infor–mation;

• understand the types and uses of epidemiology —descriptive epidemiology, surveillance epide–miology, field epidemiology, and hospital epidemiology.

MATERIALS NEEDED FOR THIS LAB

1. One nutrient agar plate per person.
2. One latex glove or small plastic sandwich bag per person.
3. One **numbered** unknown swab per person. Select swabs will have been dipped into a broth culture of *Serratia marcescens*. The others have been dipped into sterile water, which has had red food coloring added to it.
4. Beaker of disinfectant or biohazard bag.

LABORATORY PROCEDURES

1. Obtain a nutrient agar plate and divide it into five sectors, numbering them from 1 to 5.
2. Obtain a numbered swab. Be sure to record your number!
3. Place the latex glove or sandwich bag on one hand. It usually works best when all gloves/bags are on the same hand (left), with the right hand then free for swabbing. Swab the palm and fingers of your gloved hand with your numbered swab. Discard the swab in a beaker of disinfectant or a biohazard bag.
4. Shake hands, using your gloved hand, with a classmate. Be sure that your fingers come in contact with the palm of the other person's gloved hand. After shaking hands, touch your fingers on the agar in the first sector on your petri plate. Be sure to record the swab number of the person with whom you shook hands.
5. Repeat step 4 with the remaining members of your group, being sure to record each person's swab number.
6. Discard you glove or sandwich bag into a biohazard bag.
7. Incubate your plate at room temperature until the next laboratory period.
8. Following incubation, determine the presence of

Serratia marcescens (red pigmented) colonies on your plates. Record the sector(s) where they were located. Try to determine which swab(s) were contaminated and therefore the source of the "epidemic." The individual(s) with the original contaminated swab would be considered the index case(s) for the epidemic.

9. Record your results on the Laboratory Report Form.

NOTES _____

LABORATORY REPORT FORM
EXERCISE 29
EPIDEMIOLOGY

What is the purpose of this exercise?

1. Record your results in the following table:

SECTOR NUMBER	SWAB NUMBER	PRESENCE OF *Serratia marcescens*
1		
2		
3		
4		
5		

2. Compare your results with other members of your group. Which swab(s) appear to have represented the index case(s) of the epidemic?

3. What problems did you encounter in determining the index case of the epidemic?

4. What type of epidemiologic study (descriptive, surveillance, field, or hospital) would this be considered?

QUESTIONS

1. Name four portals of entry to the human body and give for each a disease-causing pathogen that would enter through that portal.

2. Often it is found that individuals who come in contact with an infected individual do not acquire the disease. What would be some explanations for this observation?

SEROLOGICAL REACTIONS

Serological reactions are specific reactions involving antigens and antibodies that occur *in vitro* or in a test tube or other artificial environment. The term *serological* is used because serum is the most convenient source of antibody. Antibody-antigen reactions are typically very specific, often approaching the specificity of enzymatic reactions. It is because of this high level of specificity that serological reactions can be used for diagnostic work.

BACKGROUND

Antigens consist of a wide variety of molecules, including proteins, most polysaccharides, nucleo–proteins, lipoproteins, and various small molecules attached to protein or polypeptide carrier molecules. These varied molecules have two properties in common. They are **immunogenic** (able to stimulate antibody formation) and exhibit specific reactivity with antibody molecules.

Antibodies are protein molecules produced by the body's immune system in reaction to exposure to a foreign or "non-self" antigen. They are characterized by containing at least two antigen binding sites. This ability to combine with two identical antigens forms the basis of many serological tests.

The interaction between antigen and antibody is usually highly specific. This means that if you know the identity of one component of an antigen-antibody reaction (either the antigen or the antibody), the other component can be detected with a very high level of sensitivity and reliability. If a reaction occurs when antibody is added to an unknown mixture, it is virtually certain that the antigen was present. When the term **homologues** is used to describe an antibody or an antigen, it indicates that the antigen and antibody can react with each other. That is, an antigen will react with its homologous antibody and, barring cross–reactions, only with its homologous antibody. Similarly, an antibody will react only with its homologous antigen.

There are many types of *in vitro* antigen-antibody reactions that can be used for the determination of the identity of a pathogenic organism or the presence of specific antibodies. These include agglutination reactions, precipitation reactions, fluorescent antibody reactions, neutralization reactions, and complement-fixation

reactions. They may be used to detect the presence of either antibody or antigen, depending upon the specific test being performed. For example, if you want to determine if a patient has been exposed to a certain virus (or bacterium or rickettsia), you could mix the patient's serum with a known sample of antigen (the organism or a component of the organism's surface). If a reaction occurs, the antibody is present in your patient's serum. Conversely, an unknown virus can be tested against known antibodies. If a reaction occurs, the identity of the organism can be determined. The various types of *in vitro* serological reactions that are commonly used in diagnostic laboratories vary in their sensitivity and the ease with which they can be performed. In this exercise we will examine several applications of the **agglutination reaction**.

Agglutination occurs when a particular antigen—such as cell-wall components, flagella or capsules—reacts with its homologous antibody and agglutinates or clumps (see Figure 30.1). Characteristically, the antigen is **polyvalent** (containing many antigenic sites), while the antibody is in most cases bivalent (able to combine with two identical antigens).

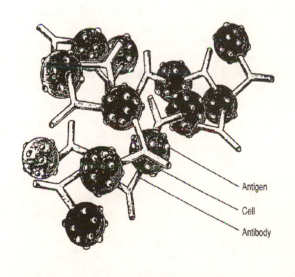

Figure 30.1   Agglutination reaction.

When the two are mixed, increasingly large complexes of antigen-antibody are formed that appear to clump together. The agglutination complexes form because the bivalent antibody (Ab) molecule reacts with two particulate antigens (Ag). Each antigen, because it is polyvalent, can also react with several antibody molecules. The repeating complex eventually gets so large that the particles appear as visible clumps (see Figure 30.2)

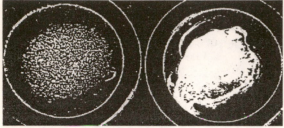

Figure 30.2   Positive (left) and negative (right) agglutination reactions.

Agglutination tests can be performed either on a slide or in a test tube. In a slide agglutination test, a drop of bacterial suspension might be combined with a drop of known antiserum for the rapid determination of homology. This can enable rapid preliminary identification of the pathogen. In other cases, a drop of the patient's serum can be combined with a known, and usually killed, suspension of bacteria to determine the presence of homologous antibodies in the patient. This usually indicates that the patient has been exposed to the specific pathogen, but does not necessarily indicate the source of a current infection.

To determine the exact level of antibody in serum, an agglutination titration must be performed. The antibody titer is an estimation of the concentration of antibody in the patient's serum. If an increase in titer is observed in successive tests, it can be assumed that the infection is caused by the test organism. The antibody titer rises relatively slowly in the early stages of an infection, rising to a maximum level and then falling off to a low level. Following a secondary exposure to the same antigen (an **anamnestic** or **memory response**), the increase in antibody titer occurs much more rapidly and to a higher level.

Because of the rapidity and ease of performing agglutination reactions, passive agglutination methods have been developed. In passive reactions, non-cellular antigens or antibodies are converted into a cellular or particulate form. It may involve the utilization of sheep red blood cells that have been treated to enable their absorption of specific antigens or the coating of latex beads with either antigen or antibody. These techniques have become the basis of many of the rapid clinical identification tests.

Modern technology has also enabled the development of highly specific antibody molecules. Early testing for antigen identification used antibodies produced primarily by the introduction of antigens into an animal (for example, a horse), which would then produce the specific antibody. These antibodies could then be used in test systems. A definite drawback to this system was the possible impurity of the antibody testing system. **Monoclonal antibodies** are used today for many of the antigen-antibody test systems (see Figure 30.3). With monoclonal antibodies, the specific antigen is injected into a mouse, which will produce antigen-specific antibodies. The antibody-producing cells (plasma cells) in the mouse can then be harvested. These cells only produce antibodies against the specific antigen; however, they are mortal cells. They can be combined with myeloma cells (a cancer cell that multiplies rapidly and is immortal), forming a **hybridoma** or antibody-producing cell that undergoes frequent and repeated cell divisions. The antibodies produced by these cells are highly specific and provide increased sensitivity in testing systems.

SLIDE AGGLUTINATION REACTIONS:
BLOOD TYPING

Human red blood cells contain varied glycoprotein and glycolipid molecules on their cell membrane surface. These molecules have specific antigenic properties. Karl Landsteiner was the first to note that if blood cells from one person and serum from a different person were combined, agglutination would occur. We now know that the basis of this agglutination is the structure of one specific glycolipid, known as substance H, which is found on the cell's surface. This glycolipid can be naked or combined with a mucopeptide (N-acetyl-galactosamine), or a sugar (galactose). Glycolipid plus N-acetylgalactosamine is commonly referred to as the A antigen. Therefore, an individual who has this modification of substance H is said to have type A blood. If galactose had been added to the glycolipid, it is said to be the B antigen and the individual is said to have type B blood. Type AB blood is due to the presence of the N-acetylgalactosamine on some of the glycolipids and galactose on others. Type O blood is characterized by the absence of either the A or the B antigen. These blood cells would, therefore, contain only the naked glycolipid.

It must be noted that the nature of substance H on red blood cells is only one of approximately 50 surface factors that have been identified as distinguishing individual red blood cells. Because it is one of the more prevalent antigenic forms, it has become the basis for our preliminary blood screening. The Rh_D antigen is also commonly screened for. When this antigen is present, the blood is said to be Rh+; if absent, Rh−.

A patient's blood type may be determined by mixing the patient's (the unknown) red blood cells with serum know to agglutinate either Group A, Group B, or Group D (Rh+) cells. The term **hemagglutination** is commonly applied to those agglutination reactions which involve red blood cells.

Agglutination With	Indicates
Anti-A	Type A
Anti-B	Type B
Anti-A and Anti-B	Type AB
Neither Anti-A nor Anti-B	Type O
Anti-D (Anti-Rh)	Type D (Rh+)
Not Anti-D (Anti-Rh)	Type Rh−

SLIDE AGGLUTINATION REACTIONS: SEROLOGICAL IDENTIFICATION OF BACTERIAL UNKNOWNS

The traditional identification of bacterial organisms involves their overnight cultivation, followed by the performance of specific biochemical tests and staining techniques. Whereas this method tends to be quite accurate, it is time–consuming and expensive. **Serotyping**, or the determination of antigens for the purpose of identification, is a much faster and easier method. The patient's serum can be tested against commercially available antigens to determine the presence of specific antibodies; pathogens can be cultivated and isolated and then tested using commercial antisera to determine their specific identity; or a specimen from the patient can be directly tested for the presence of specific antigens. An additional benefit to serotyping can be seen when tracing the course of an epidemic. For example, over a thousand distinctive antigenic variations of *Salmonella* have been identified. By determining the exact serovar (antigenic type) of *Salmonella* involved in an infection, it is possible to determine relatedness of cases in the outbreak.

Bacterial cells have two major types of antigens on their cell surface. Both types are species, and sometimes strain, specific. These antigens are found either on the outer membrane of gram-negative cells or are associated with flagella. The antigens associated with the outer membrane are commonly referred to as somatic or "O" antigens. Do NOT confuse this with type O blood. The designation O used here comes from the German *ohne hauch* or "without spreading." This is to contrast it with the flagella or "H" antigen, which is named from the German *hauch* (spreading).

Somatic "O" antigens are heat stable polysaccharides. Each form known has been given a number from 1 to 64. It is possible to group organisms in serogroups based upon the presence of common antigens. Each member of a serogroup must have at least one antigen in common. For example, all members of Serogroup A have antigen 2.

The flagellar or "H" antigens are heat–labile proteins associated with the flagella. They are commonly designated by lower case letters (i.e., group d).

When a laboratory wishes to serotype an unknown, the technician will usually determine the serotype of both the O and H antigens present on the bacterial cell surface.

In this exercise, we will use a polyvalent antibody that reacts with almost all strains of *Salmonella*. If, in the clinical laboratory, a positive reaction occurs with the polyvalent antibody, then the unknown is tested with group–specific O antibodies. Once the O antigens have been identified by slide agglutination, then the identity can be confirmed by the identification of the H antigens. This is a tube agglutination test.

FEBRILE AGGLUTINATION

The term *febrile antigens* is generally accepted as referring to bacterial suspensions representative of a number of microbial species pathogenic to man and characterized by causing a fever in their host. The most distinctive organisms in this group are members of the genera *Salmonella, Brucella, Francisella, Leptospira,* and some members of the *Rickettsia.*

The febrile agglutination test uses known antigens to determine if a patient's serum contains antibody to one or more of the antigens contained in the set. The antigens are bacterial cells that have been killed and stained. In this test, the antigen is known and one tests for the presence of homologous antibody in a patient's serum (the unknown). Known positive controls should always be run with the test.

The test is usually performed by mixing one drop of the patient's serum with one drop of each of the test antigens. If agglutination occurs, and all the controls show the expected reactions, the patient is presumed to have been exposed to the antigen. Additional testing is usually necessary to determine if the patient is harboring the infectious agent and to obtain more specific information about which strain of antigen the patient was exposed to.

One of the most commonly used agglutination reactions is the typing (A, B, O, and Rh

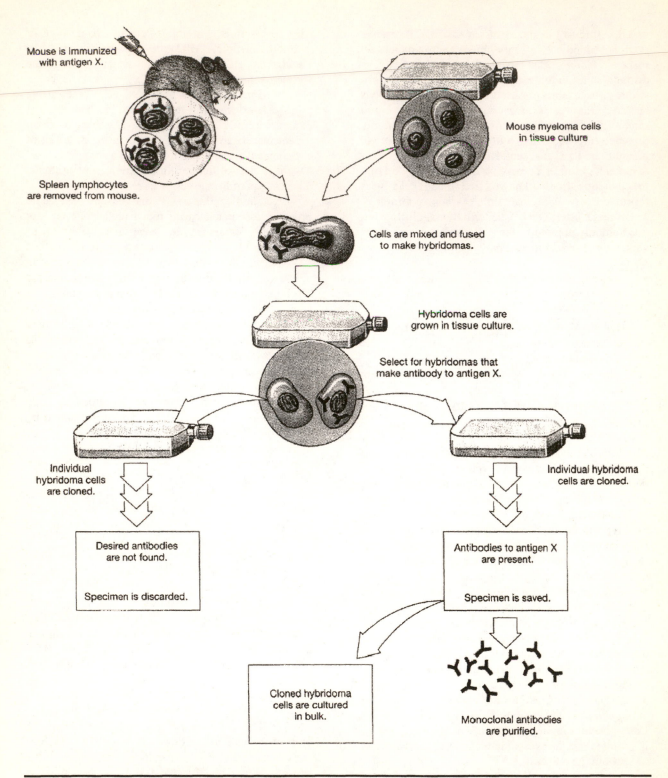

Figure 30.3   Production of monoclonal antibodies.

or D) of blood. In this exercise you will determine your blood type. You will also use agglutination reactions to serotype bacterial unknowns. (What does serotype mean?). The febrile agglutination test will be used to demonstrate how known antigen (killed bacterial cells) can be used to detect antibodies in a patient's serum. In all types of serological reactions, however, the antibody and antigen must react. These reactions are highly specific, and it is this specificity that makes the reactions so important diagnostically.

LABORATORY OBJECTIVES
Serological reactions are characteristically very specific and of diagnostic value. In this exercise you should

- understand the specificity of the antigen-antibody reaction;
- discuss the antigenic basis of blood typing;
- know how to read and interpret an agglutination reaction;
- appreciate the diagnostic value of serological reactions;
- understand why a known antigen or antibody can be used to identify the presence of antibodies or antigens.

MATERIALS NEEDED FOR THIS LAB
A. BLOOD TYPING
1. Anti-A, anti-B, and anti-Rh typing sera.
2. Glass slides.
3. Lancets and alcohol (70%) swabs or wipes.
4. Toothpicks or applicator sticks.
5. **Optional:** artificial or purchased sera may be used for blood typing in place of student blood.

B. SEROTYPING OF UNKNOWN BACTERIA
1. Polyvalent *Salmonella* O Antiserum Set and Positive Control Antigen.
2. Phenolized saline suspensions of *Salmonella* (24 hour agar slant culture of organism and at least 1.0 ml phenolized saline—0.5% phenol in 0.85% saline).
3. Glass slides or depression slides

C. FEBRILE AGGLUTINATION TEST
1. Febrile agglutination set, including positive and negative controls.
2. Unknown serum samples.
3. Agglutination plates or slides.

LABORATORY PROCEDURE
A. BLOOD TYPING
1. Obtain a glass slide and divide it into three sections with a wax pencil (see Figure 30.4).

2. Obtain samples of the three antisera (anti-A, anti-B, and anti-D). Have them ready to use.
3. Obtain a lancet. Open it and position it so that the tip does not touch the laboratory bench. It is best to leave the tip in the package until you are ready to use it.
4. Carefully clean the tip of your finger with the alcohol swab or wipe. Allow the cleansed area to air-dry.
5. Lance the fingertip by quickly stabbing it with the lancet. A quick jab is usually the least painful and most effective.
6. Allow one drop of blood to fall into each of the sections on the slide. Press a sterile wipe against the lancet wound until a clot forms and blood flow stops (about 1 minute). Dispose of the lancet and wipes in the biohazard container, as directed by your instructor.
7. Place one drop of the antisera into the appropriate section of the slide. Keep the dropper about 1 cm above the slide so the tip does not touch the blood.

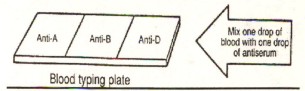

Figure 30.4   Slide-agglutination slide.

8. Using separate toothpicks for each of the three mixtures, carefully mix the blood and antiserum to obtain a homogeneous mixture. Agglu-tination should occur (if it is going to) almost immediately.
9. Discard your slide into the biohazard container.
10. Record your results on the Laboratory Report Form, Part A

NOTE: If you are using artificial or purchased sera, omit steps 1 through 5

B. SEROLOGICAL IDENTIFICATION OF BACTERIA
1. If you are not supplied with a suspension of bacteria in phenolized saline, you may prepare your own suspension. Place about 1.0 ml of the phenolized saline into a clean test tube. Transfer a loopful of the unknown bacterial culture (from an agar slant) to the phenolized saline and mix well. This process should be performed under a hood unless your instructor gives you specific instructions otherwise.

2. Divide a clean glass slide into three sections with a wax pencil (see Figure 30.4).

3. Each of the sections will contain different combinations of test and control mixtures. Usually one drop is a sufficient amount of reagent. As before, drop the test and control reagents onto the slide – do not allow droppers to touch the slide or the reagents.

4. Mix the reagents in each of the circles with separate toothpicks. Agglutination should occur almost immediately.

5. The first mixture, containing saline and the unknown bacterial suspension, should not agglutinate (negative control). The third mixture, containing the antiserum and the suspension of known positive bacteria, should agglutinate. The second mixture will agglutinate only if the organism has antigens that are homologous to one of the antibodies in the polyvalent mixture on its flagella or outer membrane.

6. Record your findings on the Laboratory Report Form, Part B.

C. FEBRILE AGGLUTINATION TEST

1. There are six bacterial antigens in the set, and each one of them must be tested against the unknown serum and against the two controls.

2. Depending upon the type of plates or slides available, mark three rows of six test areas. One row will be used for the unknown serum and the other two will be used for the controls. See Figure 30.5.

3. Place one drop of serum into each test area in the appropriate row.

4. As before, place one drop of antigen into the drop of serum. Do not allow the dropper tip to touch the plate or the drop of serum.

5. Mix each with a separate toothpick or applicator stick.

6. Agglutination should occur almost immediately; the results should be observed in about 1 minute.

7. Record your results on the Laboratory Report Form, Part C.

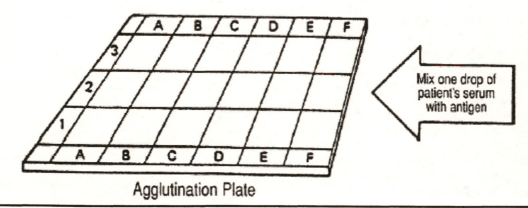

Agglutination Plate

Figure 30.5   Slide-agglutination plate.

LABORATORY REPORT FORM

EXERCISE 30
SEROLOGICAL REACTIONS

What is the purpose of this exercise?

A. BLOOD TYPING

1. Complete the following table:

ANTISERUM	HEMAGGLUTINATION
Anti–A	
Anti–B	
Anti–D	

Based on the results you got, what is your blood type?

2. Determine the blood type distribution for your class. Obtain the data from the other students

Number of students in class _____

BLOOD TYPE	NUMBER	PERCENT
Type A		
Type B		
Type AB		
Type O		
Type Rh+ (D)		
Type Rh– (d)		

B. SEROLOGICAL IDENTIFICATION OF BACTERIA

   1. Describe the agglutination reaction observed in the slide agglutination serotyping of the bacterial cultures.

   2. If this were a clinical unknown, would additional serological testing be necessary?

C. FEBRILE AGGLUTINATION TEST

   1. Fill in the following chart using your data from the febrile agglutination test

ANTIGEN	REACTION (+ OR −)
Proteus OX-19	
Brucella abortus	
Salmonella O (D)	
Salmonella H (a)	
Salmonella H (b)	
Salmonella H (d)	

QUESTIONS

   1. What type of antibodies might be found in the serum of a person with type B cells?

   2. Why are agglutination reactions useful for the identification of microorganisms?

   3. Distinguish between O and H antigens.

4. Seven individuals have been found to harbor a food-borne *Salmonella* infection that has been traced to a common source. Once this information became publicized, several additional people went into local emergency rooms, complaining of similar symptoms. How could you determine if the source of their infection is likely the same as that of the original seven individuals?

THE COLLECTION AND TRANSPORT OF CLINICAL SPECIMEN

BACKGROUND

In order to successfully determine the causative agent of a bacterial infection, a clinical specimen must be properly collected, transported, and examined. Specific methods will vary dependent upon the site being sampled, but in all cases it is important that aseptic techniques be used to prevent contamination with normal flora. Common sources of contamination are shown in the following table:

Source of Culture	Possible Contamination Sources
Urine	Urethra and perineum
Blood	Skin at site of venipuncture
Middle ear	Outer ear canal
Nasal sinus	Nasopharynx
Throat	Oral cavity and gingiva
Subcutaneous wound or abscess	Skin and mucous membranes

A. SPECIMENS

The specimens that can be examined include those that can be directly sent to the laboratory (urine, blood, cerebral spinal fluid, fecal samples, sputum, skin scrapings), as well as those that must be swabbed (abscess, ear, genital, nasopharynx, throat). The desired sample is collected in a sterile container or with a sterile swab, avoiding contact with any surrounding tissue. Once the specimen is collected it is important that it be promptly sent to the laboratory and cultured.

For any specimen to be accepted and examined by the laboratory, several criteria must be met. It must be properly labeled, promptly transported, properly sealed, and contained within a suitable transport system for the type of specimen collected.

B. TRANSPORT

To help preserve the quality of the swab specimen, to prevent the death of fastidious organisms, and the overgrowth of normal flora, it is helpful to utilize one of the several transport systems available. These systems include both aerobic and anaerobic transportation and the incorporation of one of several transport media. **Transport media** are specially designed to preserve organisms from the time they are collected until they are plated. They consist of buffered salts that prevent microbial growth, water to prevent dehydration, and oftentimes reducing agents to support the survival of anaerobic organisms.

Any specimen suspected of containing *Shigella* species must be processed immediately. If this is not possible, the transport system should be refrigerated until culturing is possible. If *Neisseria gonorrhoeae*, *Neisseria meningitides*, or *Haemophilus influenza* are suspected, the sample must not be refrigerated, as these organisms are highly susceptible to cold.

C. EXAMINATION

Once the specimen is received by the laboratory, it will be microscopically examined and cultured on suitable media as determined by the source of the inoculum and the suspected infectious agent. Processing of specimen for bacterial agents will include the utilization of Gram staining, enrichment broths, blood agar, chocolate agar, MacConkey agar, EMB agar, phenylethyl alcohol agar, Hektoen enteric agar, or Salmonella-Shigella agar. What type of infectious agent would each of the above media be used to detect?

LABORATORY OBJECTIVES

We will be utilizing a simulated wound to examine the effects of transport on successful recovery of pathogens from cultures. This will enable us to

- understand the importance of proper collection and transport of microbial specimen to the laboratory;
- evaluate the factors during transport that might alter the results of microbial testing
- evaluate the usage of aerobic and anaerobic transport systems and transport media.

MATERIALS NEEDED FOR THIS LAB

1. Twenty-four hour cultures of
   Clostridium sporogenes (thioglycollate broth)
   Staphylococcus epidermidis (tryptic soy broth).
2. Sterile test tubes (2).
3. Sterile Pasteur pipette and bulb.
4. TSA plates (6).

5. Calcium alginate swab (2).
6. Aerobic transport system (BBL CultureSwab®).
7. Anaerobic transport system (BBL Vacutainer Anaerobic Specimen Collector®).
8. GasPak® anaerobic jar with catalyst, indicator, and gas generator envelope.

LABORATORY PROCEDURE

1. Prepare mixed culture of *Clostridium sporogenes* and *Staphylococcus epidermidis* by transferring 5 to 10 drops of each culture to a sterile test tube. Use a separate sterile Pasteur pipette for each transfer.

2. Using a sterile calcium alginate swab, dampen the swab in the mixed culture. Ring the inside of the tube with the swab to avoid excess moisture from dripping on to your plates. Inoculate two TSA plates, utilizing the swab for the first section and your inoculating loop to complete the streak plate.

3. Incubate one labeled plate at 37°C under aerobic conditions. Place the second labeled plate in the GasPak and incubate it anaerobically at 37°C until the next laboratory period.

4. Moisten a second swab in the broth culture. This swab is to be placed in a sterile test tube and incubated at room temperature until the next laboratory period.

5. Following the directions for the aerobic transport system, moisten the swab in the broth culture. Replace the swab in the transport tube and store at room temperature until the next laboratory period.

6. Following the directions for the anaerobic transport system, moisten the swab in the broth culture. Replace the swab in the transport tube, follow the directions for generating the anaerobic environment and store at room temperature until the next laboratory period.

7. Examine the TSA plates incubated aerobically and anaerobically. Perform Gram stains on isolated colonies to determine identity of organism(s). Record your results on the Laboratory Report Form. Save the plates in the refrigerator until the next laboratory period.

8. Using alcohol–sterilized forceps, remove the swab from the sterile test tube stored at room temperature. Inoculate two TSA plates, using the swab for the first section on each and the loop to complete the streak plate. Incubate one labeled plate aerobically at 37°C and the second plate in the GasPak at 37°C until the next laboratory period.

9. Using the swab from the aerobic transport system, inoculate one TSA plate, using the swab for the first section of the streak plate and your inoculating loop to complete the streak plate. Incubate aerobically at 37°C until the next laboratory period.

10. Using the swab from the anaerobic transport system, inoculate one TSA plate, using the swab for the first section of the streak plate and your inoculating loop to complete the streak plate. Incubate anaerobically in the GasPak at 37°C until the next laboratory period.

11. Following incubation, examine the aerobic and anaerobically incubated plates. Perform Gram stains on isolated colonies to determine identity of organism(s) present. Compare these plates with the plates from the first day. Record your results on the Laboratory Report Form.

LABORATORY REPORT FORM

EXERCISE 31
THE COLLECTION AND TRANSPORT OF CLINICAL SPECIMEN

What is the purpose of this exercise?

1. How can a Gram stain enable you to distinguish between *Clostridium sporogenes* and *Staphylococcus epidermidis*?

2. Complete the following, indicating whether *Clostridium sporogenes* and *Staphylococcus epidermidis* are present initially and following incubation

INCUBATION CONDITION	INITIAL INCUBATION	STORED SWAB	TRANSPORT SYSTEM
Aerobic			
Anaerobic			

3. What effect, if any, did the storage method have on the preservation of organisms from time of collection to time of culturing?

QUESTIONS

1. Give three reasons why a specimen should be transported to the laboratory and processed as quickly as possible.

   a.

   b.

   c.

2. What type of media would a properly transported sample be plated on for determining the presence of the following organisms?

   a. *Escherichia coli*

   b. *Staphylococcus aureus*

   c. *Salmonella* species

3. How important is the utilization of anaerobic incubation of wound samples?

THROAT CULTURES

BACKGROUND
The upper respiratory tract has a rich population of indigenous organisms, most of which are considered to be normal flora. It also supports the growth of several important pathogens, the most important of which include the streptococci, the staphylococci, the neisseriae, the diphtheroids, yeasts (particularly *Candida*), and enteric gram-negative rods. In addition to these, which are commonly isolated from the upper respiratory tract of persons of all ages, certain other pathogens are occasionally encountered in samples from the very young (under 5 years) and elderly (over 60 years). These latter include the genera *Haemophilus, Bordetella, Yersinia,* and *Francisella*.

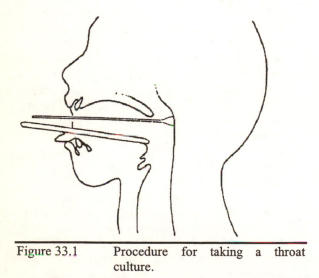

Figure 33.1   Procedure for taking a throat culture.

THROAT CULTURES: SITE TO BE SAMPLED
Throat cultures are taken from the area at the very rear of the mouth, behind the uvula, by rotating a sterile swab over the area (see Figure 33.1). Any inflamed area that might be observed should be cultured. It is very important that:
1. The culture be taken from the correct part of the oral cavity. The single most common error encountered in throat cultures is that the culture is taken from the wrong part of the mouth. The swab should be rotated over the mucosal surface behind the uvula.
2. The swab must be rotated both when the sample is taken and when the swab is used to inoculate the media. This will ensure that a large enough surface area will be sampled and that the media will be adequately inoculated.

TECHNIQUES FOR MEDIA INOCULATIONS
In general, techniques for the isolation and identification of the pathogenic forms have been well defined, and the procedures are straightforward and direct. Much of the diagnostic work involves distinguishing between the pathogenic and nonpathogenic species that comprise the normal flora. The use of special differential and selective media has greatly facilitated this task. The correct method for inoculating diagnostic media when the sample is contained on a swab is to rotate the swab on the surface of the medium so that the entire surface of the swab has had contact with the medium. You should cover an area about the size of a nickel or quarter and then use an inoculating loop to complete a streak plate (see Exercise 9).

The following selective and differential media are commonly used for routine throat cultures:
1. Blood agar
2. Chocolate agar
3. Mueller tellurite agar
4. Mannitol-salt agar

SELECTIVE AND DIFFERENTIAL MEDIA AND THEIR REACTIONS
Several of the above media will be used in this exercise. You will be asked to take a throat culture and to compare the growth obtained from it with the characteristic colonies produced by known and representative organisms. The selective and differential media you will use include:
1. **Blood agar.** Blood agar is a nutrient medium, such as trypticase soy agar, that has whole sheep

red blood cells added to it. The most common mixture uses a 5% suspension of specially washed ells in the medium. Of course, special preparatory procedures must be followed to prevent damage to the red blood cells by the heat of the melted medium. It is not uncommon to add special nutrients and/or selective agents to blood agar, depending upon the needs of the laboratory.

Bacteria that secrete hemolysins (hemolytic enzymes) are able to lyse the sheep red blood cells in the medium. The destruction of these cells will produce distinct zones of hemolysis around their colonies. They may be categorized according to the type of hemolysis they cause:

- Alpha-hemolysis: an incomplete lysis of the erythrocyte and a partial denaturation of the hemoglobin molecule to various heme products. Alpha-hemolysis results in green discoloration of the area around the colony. Microscopic examination would reveal cellular debris in the green-colored regions.

- Beta-hemolysis: a complete lysis of the erythrocyte membrane and the reduction of hemoglobin to colorless products through the release of hemolysis enzyme from the bacterial cell. Beta-hemolysis results in a completely cleared zone around the colony, and no cellular debris would be observed upon microscopic examination.

- Nonhemolytic: no visible change in the erythrocytes around the colonies. Sometimes nonhemolytic streptococci are referred to as *gamma-hemolytic*, but since there is no change in the erythrocytes, *nonhemolytic* is a more accurate description.

2. **Chocolate agar.** Chocolate agar is blood agar that has been heated to denature the proteins of the sheep red blood cells. This causes them to lyse and to turn a light brown color, not unlike that of milk chocolate (hence the name). Most formulas for chocolate agar include certain enrichment factors (such as added hemoglobin and other factors) that encourage the growth of some of the more fastidious pathogens, including *Neisseria* and *Haemophilus*.

Some bacterial hemolysins that are able to denature hemoglobin in blood agar are also able to cause changes in the color of chocolate agar. Many streptococci and lactobacilli that cause either alpha- or beta-hemolysis will cause either a greening or a decolorization of the light brown color in chocolate agar. However, if the bacteria being isolated are nonhemolytic, there will be no color change in the medium, and the colonial

morphology will be typical of the specie being isolated.

3. **Mueller tellurite agar.** Mueller tellurite agar is a selective and differential medium for the identification of *Corynebacterium diphtheriae*. This medium inhibits the growth of most cocci, non-diphtheroid bacilli and yeast and allows the differentiation of the mitis, gravis, and intermedius types of *C. diphtheriae*. The differentiation is based on the size and degree of darkening of the colonies.

4. **Mannitol-salt agar**. Mannitol-salt agar is a medium containing the sugar mannitol and 7.5% sodium chloride (salt). The salt inhibits the growth of most bacteria except the staphylococci. The mannitol differentiates between *Staphylococcus aureus* (a mannitol-fermenter) and *Staphylococcus epidermidis* (a mannitol-nonfermenter). Organisms that ferment mannitol will produce a yellow zone around the colony as a result of the production of acids during fermentation. Organisms that do not ferment mannitol will produce a deeper red color (alkaline reaction) or will not change the color at all.

DIFFERENTIATION AND IDENTIFICATION: SCREENING AND CONFIRMING TESTS

1. **Streptococci from other gram-positive cocci**. All members of the genus *Streptococcus* are catalase negative, and all other gram-positive cocci are catalase positive. The catalase test is performed by either placing one drop of hydrogen peroxide on a colony or by picking up an isolated colony from a plate and placing it in a drop of hydrogen peroxide on a microscope slide (see Exercise 16). Remember, you do not want to perform the catalase test on colonies on blood agar (why?). Catalase produced by positive organisms will cause the hydrogen peroxide to be broken down to water and oxygen, resulting in vigorous bubbling. Catalase-negative organisms do not cause bubbling.

2. **Streptococci from micrococci.** All of the staphylococci are facultative, while most of the common micrococci are obligate aerobes. The two genera can be differentiated by determining their growth characteristics in thioglycollate broth (see Exercise 11). The staphylococci will grow throughout the media, whereas the micrococci will only grow at the top. In addition, the staphylococci are usually oxidase negative; the micrococci are usually oxidase positive.

3. **Staphylococcus aureus from Staphylococcus epidermidis**. Any one of the following tests can be used to differentiate between *S. aureus* and *S. epidermidis*. The coagulase test, described in Exercise 16, is considered the most important and is the one most frequently used.

Test	S. aureus	S. epidermidis
Coagulase	+	−
Acid from mannitol	+	−
Liquefies gelatin	+	−
DNase positive	+	−

4. **Alpha-hemolytic streptococci.** Many of the alpha-hemolytic streptococci are part of the normal flora. However, many of them can be pathogenic when they are introduced into other parts of the body. One of them, *Streptococcus pneumoniae*, is the causative agent of bacterial pneumonia and can be a very serious pathogen. It must always be ruled out when alpha-hemolytic streptococci comprise a significant proportion of the isolates from a throat culture. Virtually all the alpha-streptococci except *Streptococcus pneumoniae* are resistant to lysis by bile-salt solutions and will grow in the presence of such salts. This characteristic is used to determine if any of the alpha-hemolytic streptococci that are cultured from the throat are *Streptococcus pneumoniae*. The sensitivity of the bacterium to bile salts can be measured by either placing one drop of bile-salt solution over a colony or by using prepared disks (Taxo P or Optochin). Sensitive organisms will dissolve in the bile solution or will not grow adjacent to the disk. The characteristic gram-positive diplococcus morphology of *Streptococcus pneumoniae* should always be confirmed. (How?)

5. **Beta-hemolytic streptococci.** The beta-hemolytic streptococci are **all** pathogenic and must be carefully identified. One of the most important differential tests is the sensitivity of group A streptococci to the antibiotic Bacitracin. This sensitivity is measured by placing a disk with the antibiotic on blood agar that has been streaked with the beta-hemolytic isolate. Further differentiation of the other groups (B, C, and D) of beta-hemolytic streptococci requires the determination of several biochemical characteristics.

   Because of the need for accurate and often rapid identification of the group A streptococci, several immunological tests are commonly used in the clinical laboratory. They do not require the growth of the organism, and most can be performed directly on the throat culture swab. The group A antigen is extracted from any cells present, and its presence is detected through agglutination with reactive latex (latex molecules which contain the specific group A antibody).

6. **Enterococci.** Many, but not all, of the enterococci are nonhemolytic. These organisms are part of the normal flora of the intestinal tract, and, like the alpha-hemolytic streptococci from the throat, they can be important pathogens when introduced into other parts of the body. The enterococci can be distinguished from all other streptococci, regardless of hemolytic characteristics, by their growth in 6.5% salt, and the hydrolysis of esculin. Because the enterococci may exhibit alpha- and beta-hemolysis, the presence of group D enterococci must be considered. These three tests are usually considered sufficient to confirm the presence of one of the enterococci, regardless of the type of hemolysis observed.

7. **Neisseriae.** Because up to half the colonies that appear from a throat culture can be members of the genus *Neisseria*, selective media must be used to isolate the two pathogenic species: *Neisseria gonorrhoeae* and *Neisseria meningitidis*. Growth on either of these media, when confirmed as being oxidase-positive, gram-negative diplococci, is considered presumptive for either of the species. Also, neither of them will grow on ordinary nutrient agar, whereas most of the nonpathogenic forms will. These growth characteristics are used as part of the diagnostic protocol for the neisseriae. Also included with the neisseriae is *Moraxella (Branhamella) catarrhalis*.

8. **Diphtheroids.** The diphtheroids, like the neisseriae, are among the most prominent isolates from healthy thoats. As with the neisseriae, identification of the pathogen *Corynebacterium diphtheriae* requires a special isolation medium and additional biochemical tests. Recognition of the typical gram-positive rod cellular morphology of the diphtheroids and the formation of metachromatic granules is critical for the correct identification of these organisms. Table 32.1 provides a dichotomous key to assist in the differentiation of the organisms commonly encountered in throat cultures.

LABORATORY OBJECTIVES
In this exercise you will obtain a throat culture and compare the growth obtained with known cultures provided to you. This exercise will require that you:

Table 32.1   Dichotomous Key—Bacteria Commonly Encountered in Throat Cultures

I. Gram-negative
    A. Grown on chocolate agar, but not on nutrient agar
        1. Ferments glucose and maltose   *Neisseriae meningitidis*
        2. Ferments glucose, but not maltose   *Neisseriae gonorrhoeae*
    B. Grown on nutrient agar, but not on Thayer-Martin agar
        1. Ferments at least one sugar   *Neisseria* sp.
        2. Does not ferment any sugar   *Moraxella (Branhamella) catarrhalis*
II. Gram-positive
    A. Catalase positive
        1. Aerobic   *Micrococcus* sp. (not *Staphylococcus*)
        2. Facultative
            a. Coagulase positive, ferments mannitol   *Staphylococcus aureus*
            b. Coagulase negative, does not ferment mannitol *Staphylococcus epidermidis*
    B. Catalase negative
        1. Grows in 6.5% salt and in medium containing bile   *Enterococcus*
        2. Alpha-hemolytic
            a. Soluble in bile solution   *Streptococcus pneumoniae*
            b. Not soluble in bile solution   *Streptococcus* (not *S. pneumoniae*)
        3. Beta-hemolytic
            a. Bacitracin sensitive   *Streptococcus* (Group A)
            b. Bacitracin resistant   *Streptococcus* (Group B, C, or D)

- take a throat culture;
- examine the growth that develops on the media used and attempt to recognize colonial characteristics typical of those media;
- acquire an understanding of the proper protocol for the identification of the organisms typically encountered in normal and abnormal throat cultures;
- acquire an understanding of the nature of the normal flora of the throat and how it is distinguished from the pathogenic forms that might colonize in that area;
- develop an understanding of the routine diagnostic tests employed to characterize the organisms used in this exercise.

MATERIALS NEEDED FOR THIS LAB
1. Media:
      Blood agar plate
      Chocolate agar plate
      Mueller tellurite agar plate
      Mannitol-salt agar plate
      NOTE: The chocolate and Mueller tellurite agars may be prepared as a bi–plate.
2. Sterile calcium alginate swabs.
3. Sterile tongue depressor.
4. Known cultures:
      Representative cultures, including some of the following, will be available either as broth cultures for your inoculation onto the designated

media, or for observation of their growth on the differential media.
      Moraxella (Branhamella) catarrhalis
      Corynebacterium diphtheriae
      Neisseria gonorrhoeae
      Neisseria meningitidis
      Staphylococcus aureus
      Staphylococcus epidermidis
      Streptococcus faecalis
      Streptococcus pneumoniae
      Streptococcus pyogenes
      Alpha-hemolytic *Streptococcus*
5. Optochin differentiation disks.
6. Bacitracin differentiation disks.
7. Reagents to complete testing:
      Hydrogen peroxide
      Coagulase plasma
      Oxidase reagent.

LABORATORY PROCEDURE
1. Obtain the following media:
    a. For each throat culture—one each of the following:
        Blood agar plate
        Chocolate agar plate
        Mueller tellurite agar
        Mannitol-salt agar
    b. One blood agar plate for each known streptococci
    c. One chocolate agar plate for each known neisseriae assigned

d. One Mueller tellurite agar plate for each known neisseriae assigned

e. One mannitol-salt agar for each known staphylococci assigned.

2. Prepare streak plates of assigned known organisms on the appropriate agar.

   NOTE: Your laboratory instructor may provide you with demonstration plates of the known organisms as many of them are pathogenic.

3. Using a sterile calcium alginate swab, culture the nasopharyngeal area. Be sure to rotate the swab in the tonsilar area, carefully avoiding the uvula (see Figure 32.1).

4. Referring to Figure 32.2, prepare streak plates on each of the designated media. Use the throat swab for the primary inoculation and your sterilized inoculating loop for isolation. Dispose of the swab as indicated by your instructor.

5. On the blood agar plate, place an Optochin differentiation disk on one side of your primary streak and a Bacitracin differentiation disk on the other end of the primary streak (see Figure 33.2).

6. Incubate all plates at 37°C for 24 to 48 hours.

7. Examine all plates and record the colony morphology observed. Be sure to note any reaction to the Optochin and Bacitracin disks.

8. Prepare Gram stains of representative colonies.

9. Perform those additional tests needed to aid in the identification of your isolates. Refer to the dichotomous key in Table 32.1 to assist you.

10. To perform the catalase test: Using your sterile loop, place some cells from an isolated colony on a clean glass slide. Add a few drops of hydrogen peroxide to the cells. If bubbles form, the organism is catalase positive. Place the slide in a beaker of disinfectant to kill the organisms.

11. To perform the oxidase test: Place a strip of filter paper on a glass slide. Moisten it with oxidase reagent. Using your sterile loop, rub some cells from your isolated colony on the filter paper. Observe for the production of a dark pink to purple color indicating the organism is oxidase positive. Dispose of the filter paper as indicated by your instructor.

12. Record all results on the Laboratory Report Form.

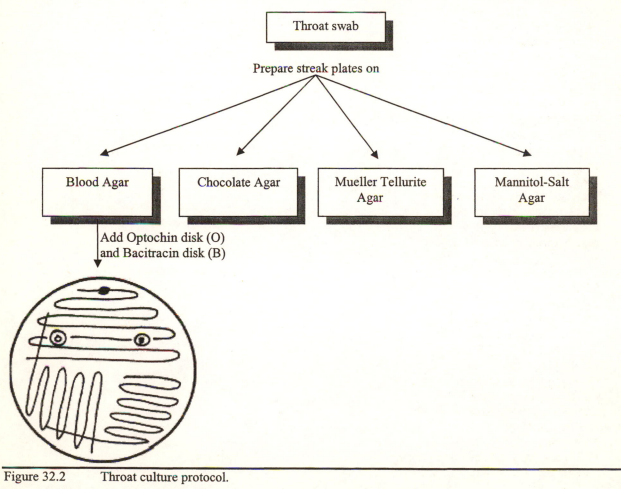

Figure 32.2   Throat culture protocol.

NOTES

LABORATOARY REPORT FORM

EXERCISE 32
THROAT CULTURES

What is the purpose of this exercise?

1. Complete the following for the known organisms tested:

ORGANISM	MORPHOLOGY	GRAM REACTION	MEDIA USED	COLONY DESCRIPTION	OTHER TEST RESULTS

2. Throat culture results

    a. Blood agar

	ISOLATE 1	ISOLATE 2
Colony Description		
Microscopic Observation		
Hemolysis Type		
Optochin Test		
Bacitracin Test		
Catalase Reaction		
Oxidase Reaction		

Which known organism(s) do the blood agar isolates appear to resemble?

b. Chocolate agar

	ISOLATE 1	ISOLATE 2
Colony Description		
Microscopic Observation		
Additional Test Results		

Which known organism(s) do the chocolate agar isolates appear to resemble?

c. Mueller tellurite agar

	ISOLATE 1	ISOLATE 2
Colony Description		
Microscopic Observation		
Additional Test Results		

Which known organism(s) do the Mueller tellurite agar isolates appear to resemble?

d. Mannitol-salt agar

	ISOLATE 1	ISOLATE 2
Colony Description		
Fermentation of Mannitol		
Microscopic Observation		
Additional Test Results		

Which known organism(s) do the mannitol-salt agar isolates appear to resemble?

QUESTIONS

1. Distinguish between alpha- and beta-hemolysis.

2. How would you differentiate between *Staphylococcus aureus* and *Streptococcus pneumoniae?*

3. Why is it important to be able to rapidly differentiate between *Staphylococcus aureus* and *Staphylococcus epidermidis?*

4. You have observed the following results for a throat culture isolate. What organism is it?
    Gram-positive
    Diplo- and streptococci
    Alpha-hemolytic
    Optochin positive
    Catalase negative

URINARY TRACT CULTURES

BACKGROUND

Urine from a healthy person, when carefully taken as a clean midstream or catheterized sample, will be found to contain remarkably few bacteria. One of the indicators of a pathological condition in the urinary tract is the presence of more than 100,000 (10^5) bacteria/ml of urine. Most laboratory protocols for the analysis of a urine culture call for some kind of quantitative measure of the bacterial density and identification of the significant potential pathogens.

ORGANISMS FREQUENTLY FOUND IN URINE

The most commonly isolated bacteria from urine cultures include the gram-negative enteric bacteria, the streptococci, and the staphylococci. *Neisseria*, *Lactobacillus*, and yeast are also often isolated. The importance of obtaining a good, clean, midstream sample cannot be overemphasized. Failure to carefully cleanse the area or to take a proper midstream sample could result in a sample that is contaminated with normal flora of the skin and associated genital organs (*Staphylococcus epidermidis*, *Lactobacillus*, and diphtheroids) and could result in a course of therapy that is not necessary.

1. The lactobacilli are slender, gram-positive rods that produce lactic acid as a product of fermentation. They are catalase negative and are part of the normal flora of the female urogenital tract, particularly the vagina.
2. Yeasts can be identified easily by their typical large size (they are eukaryotic), their characteristic formation of daughter cells by budding, and occasionally by their "yeasty" odor. Since several of the yeasts can grow on Thayer–Martin agar and do produce a weak oxidase reaction, it is imperative to do a Gram stain on all Thayer-Martin colonies.
3. The gram-negative bacilli that infect the lower urogenital tract are usually introduced from the intestinal tract. We will cover this group in detail in the next exercise.

SAMPLE COLLECTION

A clean midstream urine sample is one that is not contaminated with normal flora of the lower urogenital tract or the external genitalia. It must be carefully obtained. The patient is asked to cleanse the external genitalia and to void a portion of urine before collecting the sample. The initial voiding is necessary to rinse bacteria and cellular debris from the urethra.

SEMI-QUANTITATIVE (CALIBRATED LOOP) ESTIMATION OF NUMBERS OF BACTERIA IN URINE

The traditional method of determining the bacterial count is to perform a standard plate count on the urine sample (see Exercise 22). This is accomplished by making pour plates of a series of tenfold serial dilutions of the urine. Pour plates, however, are not suitable for rapid detection of pathogens; it is also necessary to make isolation streak plates of the sample on differential or selective media. The sample may either be streaked out directly or after mild centrifugation to concentrate the cells in a smaller volume of urine.

A more commonly used alternative procedure to the pour plate is a semi-quantitative technique that allows for the simultaneous estimation of numbers and the direct isolation of pathogens on diagnostic and selective media. This alternative technique is known as the *calibrated-loop procedure*. A calibrated loop has been carefully calibrated to hold a drop of urine of a specific volume, usually 0.01 or 0.001 ml. The loop is dipped into the urine sample and a single streak is made down the center of the plate being used for isolation. The streak is then cross-streaked for isolation of any bacteria that may have been deposited. Any medium can be used, allowing simultaneous isolation of pathogens and estimation of numbers.

Studies have established that when the loops are properly calibrated, the numbers of bacteria in the urine are proportional to the extent of growth down the center streak. If the growth extends more than three–quarters the distance, the urine is considered to have more than 100,000 bacteria/ml and is viewed as evidence of a pathogenic condition. When this occurs, the dominant organism, as well as any suspected pathogens, must be isolated and identified. A more precise estimate can be obtained by counting the colonies that appear on the streak and multiplying that number by 1,000 (if a 0.001 ml calibrated loop was used) to give the number of bacteria/ml. (What would you have to multiply your count by if a 0.01 ml calibrated loop were used?)

An even more rapid method has been developed by Wampole Laboratories. It is the Bacturcult system—a sterile, disposable, plastic tube coated with special nutrient-indicator medium for the rapid detection of bacteriuria and presumptive identification of the causative agent (see Figure 33.1). A urine sample is incubated in the Bacturcult. Following incubation, a counting strip is placed around the tube, and the number of colonies within the circular area is determined. The results are interpreted as shown in Table 33.1.

Figure 33.1   Bacturcult

Table 33.1   Bacturcult Results

Average Number of Colonies Within the Circle	Approximate Number of Bacteria/ml	Diagnostic Significance*
Fewer than 25	Fewer than 25,000	Negative bacteriuria
25 to 50	25,000 to 100,000	Borderline**
More than 50	Greater than 100,000	Positive bacteriuria

*The final diagnosis of urinary tract infection is dependent upon the clinician's judgment.
**Additional testing recommended.
Source: Wampole Laboratories

The presumptive identification of the bacteria is dependent upon the color change that occurs in the Bacturcult due to the action of the organisms on the lactose, urea, and phenol red present in the medium. This enable the differentiation of the organisms into three groups:

Group I – *E. coli, Citrobacter, Enterobacter:* yellow
Group II – *Klebsiella pneumoniae, Staphylococci, Streptococci*: rose to orange.
Group III – *Proteus, Pseudomonas*: purplish-red (magenta)

Mixed cultures do not necessarily produce clear results, and additional testing should be performed. The colonies present in the Bacturcult can be used for inoculation on additional differential media.

MEDIA USED AND THEIR REACTIONS

1. **Blood agar.** Many of the streptococci isolated will be enterococcus and will appear to be nonhemolytic (gamma-hemolysis). If the sample contains large numbers of other bacteria, the detection of the streptococci may be difficult. Refer to Exercise 30 for a complete description.

of the reactions on blood, chocolate, and mannitol-salt agars

2. **Thayer-Martin agar.** The chocolate agars, including Thayer-Martin agar, are used to detect pathogenic neisseriae. Although most bacteria can grow well on chocolate agar, the selectivity of Thayer-Martin agar will ensure isolation of *N. gonorrhoeae,* should it be present.

3. **Mannitol-salt agar.** This agar is occasionally used for the detection of *Staphylococcus aureus,* although when *Staphylococcus aureus* is present it can usually be detected on the blood agar streaks.

4. **MacConkey's agar.** MacConkey's agar is a medium that is used for the selection and differentiation of the enteric gram-negative rods associated with the intestinal tract. It contains crystal violet to suppress the growth of gram-positive bacteria. Other components of the medium, including the sugar lactose and a pH indicator (neutral red: red at 6.8, yellow at 8.0), all differentiation between lactose-fermenting and lactose-nonfermenting gram-negative rods. Lactose-positive bacteria will produce red colonies with a distinctly darker red zone around the colony (due to the color change of the pH indicator), while lactose-non-fermenting bacteria will produce colonies that are light pink or colorless after a 24–hour incubation. Whenever gram-negative bacilli comprise a significant proportion of the isolated colonies, identification is required—especially if the total count exceeds 100,000/ml.

5. **EMB agar.** EMB (eosin–methylene blue) agar is a selective and differential medium for the detection of the gram-negative enteric bacteria. Organisms utilizing lactose and/or sucrose will appear as blue-black colonies with a greenish metallic sheen, whereas coliform organisms that do not utilize the lactose and/or sucrose form mucoid, pink colonies. The presence of a green

metallic sheen on EMB agar is especially diagnostic for the presence of *Escherichia coli*.

LABORATORY OBJECTIVES
This laboratory exercise will introduce you to new procedures that are commonly used in the bacteriological examination of urine, including quantitative techniques for the estimation of the numbers of bacteria in urine. To complete this exercise you will:

- understand and explain what a serial dilution is and what it is used for;
- know what a calibrated loop is and what its applications are;
- obtain an understanding of the proper protocol for bacteriological analysis of urine samples;
- discuss the importance of a properly obtained urine sample;
- achieve some appreciation for the normal and pathogenic flora of the lower urogenital tract;
- obtain an understanding of the application of rapid indicator systems for presumptive screening of urine samples.

MATERIALS NEEDED FOR THIS LAB
NOTE: You may not be required to complete all of the procedures listed in the next section of this exercise. For example, your instructor may have some groups do standard plate counts and others do the calibrated loop count.

1. Urine samples. Urine samples may be provided for you. It is important that urine samples be processed as soon as possible following collection. If this is not possible, the sample should be refrigerated until ready for use.
2. Gram stain reagents.

Standard Plate Count
1. Three sterile 9.0 ml saline dilution blanks.
2. Four plate-count agar talls, melted and held at approximately 50°C.
3. Four sterile petri plates.
4. Sterile pipettes, 1.0 ml.

Calibrated-loop Count
1. Calibrated loop.
2. One plate-count agar.

Characterization of Known Organisms
1. Known cultures: Representative cultures, including some of the following, will be available either as broth cultures for your inoculation on the designated media, or for observation of their growth on the differential media.

Staphylococcus aureus
Staphylococcus epidermidis
Streptococcus faecalis
Lactobacillus sp.
Pseudomonas aeruginosa
Escherichia coli
Saccharomyces cerevisiae

2. Media (1 each per known culture assigned by your instructor:
   Blood agar
   MacConkey's agar
   Thayer-Martin agar
   Mannitol-salt agar
   EMB Agar
   NOTE: These may be prepared as bi-plates or as individual plates.
3. Reagents needed to complete screening tests:
   Hydrogen peroxide
   Coagulase plasma
   Oxidase reagent.

Identification of Urinary Tract Organism(s)
1. Media (1 per urine sample)
   Blood agar
   MacConkey's agar
   Thayer-Martin agar
   Mannitol-salt agar
   EMB agar.
2. Reagents needed to complete screening tests:
   Hydrogen peroxide
   Coagulase plasma
   Oxidase reagent.

Use of Bacturcult
1. Bacturcult culture tube (1 per urine sample).

LABORATORY PROCEDURES
Refer to Figure 33.2 for an overview of the procedures to be performed on the urine sample.
1. Perform Gram-stain of urine sample. Record the results on the Laboratory Report Form.

Standard Plate Count (Refer to flow chart in Exercise 22)
1. Obtain:
   a. Four sterile petri plates and label them "undiluted," "1:10," "1:100," and "1: 1000."
   b. Three 9.0-ml dilution blanks containing sterile saline. Label them "1:10," "1:100," and "1:1000."
   c. Four plate-count agar talls, melted and held at 50°C.
2. With a sterile 1.0-ml pipette, transfer 1.0 ml of urine into the first dilution blank and transfer 1.0

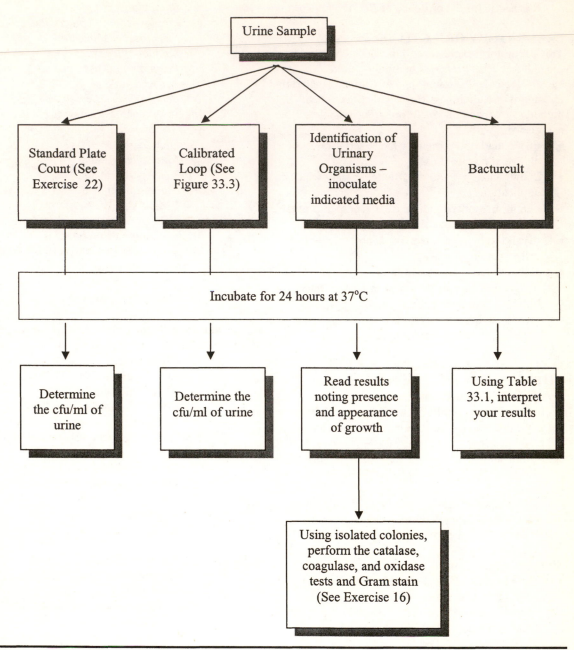

Figure 33.2   Urine testing procedure.

ml of urine into the first tube of melted agar ("undiluted").

3. Dispose of the pipette as indicated by your instructor. Do not use a pipette more than once.

4. Mix the agar by gently rolling the tube between your hands. Do not shake vigorously. Pour the inoculated media into the first plate and allow it to solidify.

5. Mix the inoculated dilution blank by rolling it between your hands. Transfer 1.0 ml from this tube into the second dilution blank and 1.0 ml into the second melted agar tall.

6. As before, mix the medium gently, pour into a petri plate, and allow to solidify.

7. Similarly, mix the second dilution blank and make the transfers to the third dilution blank and to the third melted agar tall.

8. Mix the inoculated agar gently and pour into the third petri plate. Allow the plate to solidify.

9. Mix the final dilution blank. Transfer 1.0 ml to the final melted agar tall.

10. Mix the inoculated agar gently and pour into the third petri plate. Allow the plate to solidify. You should have one plate each for undiluted urine, 1:10, 1:100, and 1:1000.

11. Incubate at 37°C for 24 hours. If you are unable to perform the count at that time, the plates should be refrigerated until your next laboratory period.

12. Determine the cfu (colony–forming units) per milliliter of urine. Remember to take into account the dilution of your urine sample.

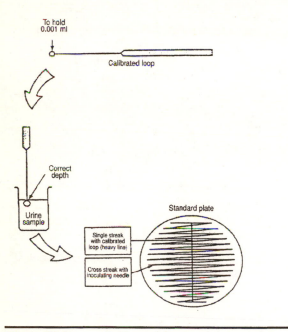

Figure 33.3   Calibrated-loop streak procedure.

Calibrated Loop Count (Follow the flow diagram in Figure 33.3)

1. Carefully clean the loop following any special instructions for cleaning and flaming provided by your instructor.

2. Dip the loop into the urine sample so that the loop is just below the surface of the sample. No part of the stem should be in the urine (urine adhering to the stem will cause the counts to come out higher than they should).

3. When you remove the loop from the urine, the loop will be filled with a droplet of sample. If the loop has been properly calibrated and correctly handled, it will contain 0.001 ml.

4. Make a single streak down the center of the plate by touching the filled loop to the top center and quickly sliding the loop down the bottom center of the plate. Cross-streak over the first streak to

isolate colonies. Use your regular loop or needle for this isolation streak.

5. Incubate at 37°C for 24 hours. If you are unable to perform the count at that time, the plates should be refrigerated until your next laboratory period.

6. Determine the number of bacteria per milliliter of urine. Remember to take into account the size of the original inoculum. Compare the results obtained with the calibrated loop with those of the standard plate count.

7. Record your results on the Laboratory Report Form.

Characterization of Known Organisms

1. Inoculate each known organism assigned to you by your instructor on each of the following media. Be sure to streak each for isolation.
   Blood agar
   MacConkey agar
   Thayer-Martin agar
   Mannitol-Salt agar
   EMB agar

2. Incubate all plates at 37°C until the next laboratory period.

3. Perform the catalase, coagulase, and oxidase tests on isolated colonies. Refer to Exercise 16 as necessary for the required procedures.

4. Complete the table in the Laboratory Report Form for each known organism culture tested.

Identification of Urinary Tract Organism(s)

1. Inoculate the urine sample on each of the following media. Be sure to streak each for isolation.
   Blood agar
   MacConkey agar
   Thayer-Martin agar
   Mannitol-salt agar
   EMB agar

2. Incubate all plates at 37°C until the next laboratory period.

3. Perform the catalase, coagulase, and oxidase tests on isolated colonies. Refer to Exercise 16 as necessary for the required procedures.

4. Complete the table on the Laboratory Report Form. Use Table 33.2 and the results of the known organisms to assist with your identification

Use of Bacturcult

1. Label the Bacturcult tube.

2. Fill the tube almost to the top with the urine sample.

3. Immediately pour the urine out of the tube, allowing the fluid to drain for several seconds.

Place the lid securely back on the Bacturcult tube.

4. Loosen the cap on the Bacturcult tube by turning it counterclockwise for half a turn.

5. Incubate the tube with the cap down at 37°C for 24 hours. If you cannot read the results at 24 hours, refrigerate the Bacturcult until your next laboratory period.

6. Place the counting strip around the tube over an area of even colony growth and count the number of colonies within the circle. Using Table 33.1, determine the approximate count per milliliter and the diagnostic significance.

7. Observe the color of the medium. Which presumptive group would this indicate?

8. Record the results on the Laboratory Report Form.

Table 33.2   Dichotomous Key—Bacteria Commonly Isolated from Urine

I. Gram-positive bacteria	
A. Catalase positive	
1. Coagulase positive	*Staphylococcus aureus*
2. Coagulase negative	*Staphylococcus epidermidis*
B. Catalase negative	
1. Cocci	*Streptococcus faecalis*
2. Bacilli	*Lactobacillus* sp.
II. Gram-negative bacteria	
A. Oxidase positive	*Pseudomonas aeruginosa*
B. Oxidase negative	*Escherichia coli*
III. Typical yeast morphology	*Saccharomyces cerevisiae*

Note: This key is limited to those tests necessary for the identification of the bacteria in this exercise. Additional organisms would require additional tests. Also, a good microbiologist would insist on some confirmatory tests for each isolate.

LABORATORY REPORT FORM

EXERCISE 33
URINARY TRACT CULTURES

What is the purpose of this exercise?

1. Describe the morphology and Gram-reaction of the organisms seen in the Gram-stain of the original urine sample.

2. Complete the results of the standard plate count and calibrated-loop count for the urine samples.

Sample Source	cfu/ml of Urine	
	Plate Count	Calibrated-Loop Count

3. Complete the following for the known organisms observed:

Organism	Observed Growth (presence or absence, appearance)				
	Blood Agar	MacConkey Agar	Thayer-Martin Agar	Mannitol-Salt Agar	EMB Agar
Staphylococcus aureus					
Staphylococcus epidermidis					
Streptococcus faecalis					
Lactobacillus sp					
Pseudomonas aeruginosa					
Escherichia coli					
Saccharomyces cerevisiae					

Organism	Test Reaction			Morphology	Gram Reaction
	Catalase	Coagulase	Oxidase		
Staphylococcus aureus					
Staphylococcus epidermidis					
Streptococcus faecalis					
Lactobacillus sp.					
Pseudomonas aeruginosa					
Escherichia coli					
Saccharomyces cerevisiae					

4. Complete the following for the urine culture tested:

Urine Sample	Observed Growth (presence or absence, appearance)				
	Blood Agar	MacConkey Agar	Thayer-Martin Agar	Mannitol-Salt Agar	EMB Agar

Urine Sample	Test Reaction			Morphology	Gram Reaction
	Catalase	Coagulase	Oxidase		

The identity of the organism(s) in the urine sample is:

5. Complete the following with the Bacturcult results.

Urine Sample	Number of Colonies in Circle	Approximate Number of Bacteria/ml	Diagnostic Significance	Color of Medium	Presumptive Group

How do the results from the Bacturcult test compare with those achieved with direct count and traditional media/tests?

QUESTIONS

1. Explain how a clean, midstream, or clean-catch urine sample is obtained.

2. What is bacteriuria?

3. List four commonly encountered pathogens associated with lower urinary–genital-tract infections.

    a.

    b.

    c.

    d.

4. Why are relatively few microorganisms found in the urine of a healthy individual?

5. Why are females more prone to the development of urinary tract infections than males?

6. What would be the effect of a urine sample being left in a patient's room for several hours before being transported to the laboratory?

GASTROINTESTINAL TRACT CULTURES

Bacterial diseases of the gastrointestinal tract can take many forms—from moderately discomforting "stomach flu" to severe diarrhea that could result in life-threatening loss of fluids and electrolytes. Almost every genus of the gram-negative enteric bacteria have species that can be pathogenic. Two of the more prevalent ones are *Salmonella* and *Shigella*, but even some strains of *Escherichia coli* can cause severe hemorrhagic colitis. Of course, all of these bacteria, whether usually considered pathogenic or not, are to be considered opportunistic pathogens that will result in severe bacterial infections when introduced to parts of the body with which they are not usually associated.

BACKGROUND

Some nonenteric gram-negative rods, such as certain species of *Pseudomonas* and V*ibrio*, are frequently encountered with the enteric bacteria. One of them, *Pseudomonas aeruginosa,* is usually thought of as nonpathogenic, but can be a very serious pathogen in immunologically compromised hosts (for example, AIDS patients and the elderly) and in burn patients. *Ps. aeruginosa* is metabolically very versatile and resistant to many of the commonly used antibiotics. It produces a characteristic blue-green pigment in many kinds of growth media. *Vibrio* is a genus of gram-negative curved rods. Most of the species in this genus are saprophytes, especially in marine environments. Two notable exceptions include *V. cholerae*, which causes cholera, and *V. para–haemolyticus*, a species associated with food poisoning involving seafood. Both *Pseudomonas* and *Vibrio* are oxidase positive and lactose negative.

The intestinal tract also provides an ideal growth environment for many other kinds of bacteria, including several kinds of enterococci and many anaerobes. Such a large and diverse assembly of bacteria presents some unique problems when it becomes necessary to detect pathogenic species that might be present in very small numbers. Several procedures using highly selective media and isolation or enrichment media have been developed to help resolve this problem.

STRATEGIES FOR STOOL CULTURES

Enrichment cultures. When some members of a bacterial population are present in only a very small proportion of that population, it is necessary to use enrichment media for their isolation. The medium chosen is one that contains selective inhibitors that will inhibit the growth of all bacteria except the species to be isolated. The objective of the procedure is to increase the likelihood of isolating the pathogenic species on a streak plate—something that would be very difficult without first enriching the population for the pathogen, as normal flora would tend to overgrow the plate.

Selective media. Most of the media used for isolation of the enteric bacteria contain inhibitors to suppress the growth of nonenteric organisms. The two most commonly used inhibitors are the bacteriostatic dyes and bile salts. The dyes inhibit the growth of gram-positive bacteria, and the bile inhibits those bacteria that are not normally found in the intestinal tract. (Why?)

Differential media. Many of the differential media used to identify the gram-negative rods do so on the basis of fermentation reactions. When acid is produced, a pH indicator in the medium changes color; a good rule of thumb to follow is that if the colony is a color different from the medium, it ferments the sugar. Most of these media use lactose or sucrose as the indicator sugar because virtually none of the pathogenic forms (with one or two notable exceptions) ferment one or the other of these sugars. Therefore, any colony that remains the same color as the medium should be considered suspect and identified. Several media also use additional selective or differential devices. Most of the media commonly used are both selective and differential.

Multitest media. In addition to the plated media, there are several tubed media that have differential properties and can be used to perform more than one test on an isolate. One of the most common is triple sugar iron agar (TSIA), which contains three sugars and a pH indicator. The sugars are lactose, sucrose, and glucose, but the lactose and sucrose concentration is ten times (10×) higher than the glucose concentration.

When this medium is correctly inoculated, bacteria that ferment glucose only will cause the medium to turn yellow in the butt of the tube and red on the slant. (Why?) Bacteria that ferment either lactose or sucrose will cause a yellow color to develop throughout the tube. If gas is also produced, bubbles will appear in the agar, causing it to "break" into pieces. This medium also contains iron salts that will turn black when hydrogen sulfide is produced. The TSIA can, therefore, be used to determine fermentation reactions of three sugars and can test for the production of hydrogen sulfide.

Summary. A typical strategy for the bacterial analysis of a stool culture (or any culture that may contain gram-negative rods) is as follows:
1. Enrichment: use of a medium that will enhance the growth of the bacterium you want to isolate so that it becomes a prominent member of the population and can be more easily isolated on a streak plate.
2. Selection and differentiation: use of appropriate media that inhibit the growth of nonenteric bacteria and which differentiate between those that ferment lactose or sucrose and those that do not.
3. Biochemical identification: identification on the basis of biochemical reactions, including those determined on certain multitest media.

MEDIA USED AND THEIR REACTONS
It is especially important that you study the growth patterns and colony morphologies of the reference cultures on the media listed below. Subtle differences in color, colony margin, consistency, and so on can be very important diagnostic leads.
1. **Selenite F broth.** This is a highly selective medium that is used to enhance the growth of the genera *Salmonella* and *Shigella*. The liquid medium is inoculated with a swab saturated with the stool sample and incubated for 12 to 24 hours. Tetrathionate broth is another example of this type of enrichment medium.
2. **Alkaline peptone medium.** Alkaline peptone medium (APM) is used to enrich samples for *Vibrio*. It contains a high salt concentration (3.0%) and a high pH (about 8.0).
3. **MacConkey's agar.** This medium was discussed in Exercise 32. The combination of bile salts and bacteriostatic dyes makes the medium selective for the enteric bacteria. Lactose and pH indicators allow for differentiation between lactose-fermenting and nonfermenting bacilli.
4. **Hektoen enteric agar.** This medium is highly selective and differential. Organisms that ferment lactose will produce yellow to pink colonies while nonfermenters will be smaller and

green in color. This medium is very reliable ad is one of the most commonly used today.
5. **SS agar.** Salmonella-Shigella (SS) agar is selective for these two genera. Like most such media, the selectivity is not absolute (*Pseudomonas* grows fairly well on it) and gram stains and confirming biochemical tests are always required. *Salmonella* and *Shigella* colonies are light pink, with *Salmonella* occasionally producing a colony with dark centers. The occasional lactose fermenter that does grow on SS agar will produce a distinctly red colony.
6. **Triple sugar iron agar.** The TSIA has been discussed in Exercise 17. A yellow color indicates fermentation of one of the sugars. When only glucose is fermented, the deepest part of the medium (the butt) will be yellow, while the part of the medium exposed to air (the slant) will be red. The red color appears when glucose is depleted, forcing the bacteria to use the protein components of the medium for growth. When either or both of the other sugars is used, the entire tube will be yellow because the concentration of the sugar is great enough so it does not become depleted. As noted above, the iron will react with any hydrogen sulfide produced and will cause the formation of a black precipitate in the deep part of the butt. Another medium, Kligler's iron agar (KIA), is very similar to TSIA except that it has only two sugars (glucose and lactose).
7. **SIM agar.** Sulfur,indole,motility (SIM) agar has been discussed in Exercise 17. SIM agar is a medium that can be used to determine sulfide (S) production, indole (I) production, and motility (M). The medium is soft enough so motile bacteria can swim through it, producing a very hazy and cloudy central stab. Nonmotile bacteria produce a sharp and clear stab as they do not move away from the site of inoculation. The sulfide is detected by formation of a black precipitate, and indole is detected by the addition of Kovac's reagent.
8. **TCBS agar.** TCBS stands for thiosulfate citrate bile sucrose agar. It is highly selective for the genus *Vibrio* because of the high pH (about 8.0) and high salt content. It differentiates between sucrose-fermenting and nonfermenting bacteria. The medium can be used for the isolation of *Vibrio cholerae* (yellow colonies) and *Vibrio parahaemolyticus* (green colonies). It is often included in stool-culture protocols during the summer months when *Vibrio parahaemolyticus* is likely to be encountered.

RECOMMENDED PROTOCOL FOR STOOL CULTURES

Day 1. Inoculate at least one enrichment medium and at least three selective and differential media by streak plate. Some workers recommend that half of the plates be used for the initial streak, the other half used for streaking the enrichment cultures (see Figure 34.1) The recommended media include:

a. Enrichment media:
   Selenite F or tetrathionate broth
   Alkaline peptone medium.

b. Isolation media:
   MacConkey's agar
   TCBS agar
   Hektoen Enteric agar
   SS agar.

Day 2. Examine the plates previously inoculated and make streak plates of the enrichment culture.

Day 3. Examine all plates for the presence of lactose nonfermenting bacteria. Also examine the TCBS plates for typical *Vibrio* colonies (yellow or green). Examine all reference cultures. Initiate any biochemical identifications that are appropriate, including TSIA, SIM, and/or KIA media.

THE IMViC TST

The IMViC test is commonly used in the bacterial analysis of water to differentiate rapidly between *Escherichia coli* and *Enterobacter aerogenes*. It is not as commonly used in clinical laboratories because other screening tests have proven more useful. Nevertheless, virtually all screening tests for the gram-negative bacilli use these reactions (in addition to others) in their protocol. The biochemical tests comprising the IMViC test are indole (I), methyl red (M), Voges-Proskaur (V), and citrate (C). Note that the *i* is simply inserted as a phonetic device. See Exercises 14 and 15 for discussions of each of these test procedures. Compare the results for *Escherichia coli* and *Enterobacter aerogenes* for these tests.

LABORATORY OBJECTIVES

Stool cultures are unique because, more so than with any other type of culture, the selection and enrichment for pathogens is an important part of the procedure. In this exercise you will:

- understand the enrichment process and why it is needed.
- understand the principles involved in the use of selective and differential media.
- correctly identify lactose- and sucrose-fermenting bacteria on the appropriate media.
- understand the procedures and strategies employed in the analysis of stool cultures.

MATERIALS NEEDED FOR THIS LAB

1. Stool specimen. This may be an actual stool sample or a mixed broth sample containing common enteric organisms.

2. Known cultures: Representative cultures, including some of the following, will be available either as broth cultures for your inoculation onto designated media or for observation of their growth on the differential media.
   Escherichia coli
   Enterobacter aerogenes
   Proteus vulgaris
   Proteus mirabilis
   Pseudomonas aeruginosa
   Vibrio parahaemolyticus
   Salmonella typhimurium
   Shigella sp.

3. Media needed for each known culture assigned and for each stool sample:
   Selenite F broth
   Alkaline peptone medium
   MacConkey's agar
   TCBS agar
   Hektoen enteric agar
   SS agar.

4. For each known culture assigned and for each isolate:
   TSIA slant
   SIM agar tall.

5. Gram stain reagents.

LABORATORY PROCEDURES

1. Inoculate Selenite F broth and alkaline peptone medium with a swab saturated with the stool sample.

2. Perform primary isolation of the stool sample of the following agars:
   MacConkey's agar
   TCBS agar
   Hektoen enteric agar
   SS agar
   Use only half of each plate for the primary isolation. The other half will be used for streaking the enrichment cultures (see Figure 34.1).

3. Prepare streak plates of assigned known organism on the following agars:
   MacConkey's agar
   TCBS agar
   Hektoen enteric agar
   SS agar.

NOTE: Your laboratory instructor may provide you with demonstration plates of the known organisms.

4. Incubate the enrichment broths and agar plates for 24 hours at 37°C. If you cannot streak them for isolation or read the results at 24 hours, they should be refrigerated until your next laboratory period.
5. Record the results from the primary isolations on the agar plates.
6. Streak out the enrichment cultures on the remaining half of the agar plates.
7. Incubate at 37°C for 24 hours.
8. Observe the growth of the known cultures on the varied media.
9. Record the results for the known cultures on the Laboratory Report Form.
10. Inoculate the known cultures into TSIA and SIM agar.
11. Incubate at 37°C for 24 hours.
12. Examine all streak plates, study and compare all colony types, and determine if any lactose non–

fermenting bacterial colonies developed on the stool culture streak plates.
13. Inoculate any suspect colonies into TSIA and SIM agar.
14. Read and record the results for the known cultures in the TSIA and SIM agar.
15. Write a report summarizing your analysis of the stool cultures. Using Table 34.1, determine the possible identity of the isolated colonies. It should include:
   a. A description of the known cultures on each of the media used.
   b. A discussion on the use of the enrichment procedure.
   c. A summary of the stool culture analysis: presence or absence of potential pathogens and protocol used for identification or screening

Table 34.1   Dichotomous Key—Bacteria Commonly Isolated in Gastrointestinal Tract Cultures

I. Oxidase positive
  A. Ferments sucrose, curved rod   *Vibrio parahemolyticus*
  B. Does not ferment any sugar, produces green pigment in some media   *Pseudomonas aeruginosa*
II. Oxidase negative
  A. Ferments lactose
    1. Produces indole   *Escherichia coli*
    2. Does not produce indole   *Enterobacter aerogenes*
  B. Does not ferment lactose
    1. Hydrolyzes urea
      a. Produces indole   *Proteus vulgaris*
      b. Does not produce indole   *Proteus mirabilis*
    2. Does not hydrolyze urea
      a. Motile   *Salmonella typhimurium*
      b. Nonmotile   *Shigella* sp.

NOTE: This key is limited to those tests necessary for the identification of the bacteria used in this exercise. Additional organisms would, of course, require additional tests. Also, a good microbiologist would insist on some confirmatory tests for each isolate.

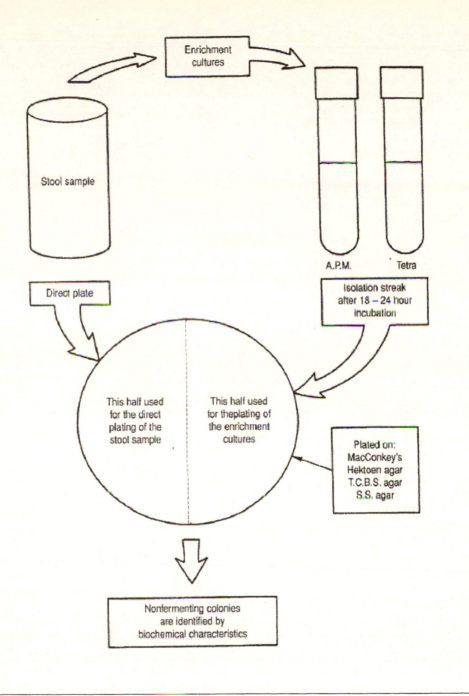

Figure 34.1   Gastrointestinal tract culture protocol.

NOTES_____

LABORATORY REPORT FORM

EXERCISE 34
GASTROINTESTINAL TRACT CULTURES

What is the purpose of this exercise?

1. Complete the following chart for the known organisms tested:

Organism	Colony Color			
	MacConkey	TCBS	Hektoen Enteric	SS
Pseudomonas aeruginosa				
Proteus vulgaris				
Proteus mirabilis				
Shigella sp.				
Escherchia coli				
Salmonella typhimurium				
Enterobacter aerogenes				
Vibrio parahaemolyticus				

2.   Complete the following chart for the known organisms tested:

Organism	TSIA and/or SIM agar				
	Color of Butt	Color of slant	H$_2$S Production	Indole Test	Motility
Pseudomonas aeruginosa					
Proteus vulgaris					
Proteus mirabilis					
Shigella sp.					
Escherichia coli					
Salmonella typhimurium					
Enterobacter aerogenes					
Vibrio parahaemolyticus					

3.   Complete the following for the stool specimen tested, describing the color of the isolated colonies:

Stool Isolate	PRIMARY ISOLATION				ENRICHMENT ISOLATION			
	MacConkey	TCBS	Hektoen Enteric	SS	MacConkey	TCBS	Hektoen Enteric	SS

4. Complete the following for each stool isolate tested:

Stool Isolate	TSIA			SIM		
	Color of Butt	Color of Slant	H$_2$S Production	H$_2$S Production	Indole	Motility

5. Based on your results above, what organisms appear to have been present in the stool sample tested?

QUESTIONS

1. Why are lactose and sucrose the two most commonly used indicator sugars in the isolation of enteric organisms?

2. What is the purpose of culture enrichment?

3.   What are bacterial endotoxins?

4.   What is meant by *enteropathic E. coli*?

LACTOBACILLUS ACTIVITY IN SALIVA

Tooth decay has been linked to bacteria found in the oral cavity. Bacteria that are part of the normal flora can form plaque and become embedded in lesions on the teeth. Through a combination of acid production by these bacteria and decomposition of tooth structure, dental caries form and increase in size.

BACKGROUND
The formation of lactic acid by bacteria belonging to the genera *Streptococcus* and *Lactococcus* has been directly linked to the formation of dental caries. These bacteria growing in saliva and on the surface of teeth ferment available carbohydrates to produce the lactic acid. Most of these bacteria are able to ferment sucrose (table sugar) and complete its conversion to lactic acid in the plaque deposits on the teeth or in the saliva of the mouth. This fermentation reaction is completed before the sucrose-containing food is completely chewed and swallowed. (Those who have inflamed gums or teeth with dental caries will feel pain as the acids produced by these bacteria irritate the lesions.) The lowered pH increases the solubility of the enamel, resulting in dental caries. The formation of plaque and the lodging of food and bacteria in tooth lesions create anaerobic conditions that enhance the acid production through fermentation. The resulting process is, therefore, self-sustaining once it begins. When these organisms produce sticky polysaccharide polymers of dextran, the lactic acid is held against the surface of the tooth as plaque, enabling their breakdown of tooth enamel and dentin. This results in increased tooth decay. It should be noted that sucrose is also important in the formation of the dextran layer.

MEDIA USED
1. **Rogosa SL agar.** Rogosa SL agar is a low pH and high-acetate medium that suppresses most of the oral flora except the lactobacilli. (Is this medium selective or differential?)
2. **Snyder test agar.** Snyder test agar provides a simple correlation between the time of acid production and the number of lactobacilli present in the oral cavity. Brom cresol green is used as a pH indicator. It turns yellow at a pH below 4.6. At this point, the organisms present have produced enough acid to decalcify the tooth enamel. When only using Snyder test agar as an

indicator of the cariogenic potential flora, the following interpretation is commonly made:

Caries Potential	Duration of Incubation		
	24 Hours	48 Hours	72 Hours
Marked	Positive	--	--
Intermediate	Negative	Positive	--
Slight	Negative	Negative	Positive
Insignificant	Negative	Negative	Negative

A positive test is considered as a change in color, as compared to an uninoculated control, with green no longer being dominant. A negative test would then be the absence of any color change, or a tube in which there is only slight deviation, with green remaining the dominant color.

EXPERIMENTAL DESIGN
Two procedures have been developed to estimate the amount of acid production of the bacteria in the oral cavity. Both procedures require that saliva be added to a selective medium and acid production measured by color changes in the medium.

One approach is to actually count the number of acid-producing bacteria present in saliva. The other approach is to assume that acid production is proportional to the number of bacteria and then to merely determine the rate at which the acid is produced. In other words, the more acid, the more bacteria.

In the Rogosa test, direct pour plate counts of the bacteria in saliva are made using Rogsa SL agar. This medium is highly selective for the genera *Streptococcus* (S) and *Lactobacillus* (L). The number of colonies that develop is assumed to be representative of the bacterial activity in the oral cavity.

As we saw earlier, in the Snyder test saliva is added directly to a tube of Snyder test agar and incubated for up to 72 hours. Based on the time it takes for the medium to turn yellow, correlations have been drawn between time and the approximate number of *Lactobacillus* bacteria/ml present in the saliva. Table 35.1 shows this correlation. Both the Snyder test and the Regosa test, in theory, measure the same thing—the amount of lactic acid produced by bacteria in saliva. In the Snyder test, the activity (production of acid) of the bacteria is measured, whereas in the Rogosa test the actual number of acid-producing bacteria in saliva is determined. Both procedures assume that there is some direct relationship between the activity or numbers measured under laboratory conditions and the actual production of acid on the surfaces of the teeth.

Table 35.1   Color Changes and Number of Bacteria/ml of saliva

Time of Color Change (Snyder Test)	Susceptibility to Dental Caries	Approximate number/ml (Rogosa Test)
< 72 hours	Low	< 100/ml
< 48 hours	High	< 10,000/ml
<24 hours	Very High	> 10,000/ml

The value of these tests lies in the relationship that exists between the production of acid by bacteria and the formation of dental caries. The accuracy and reliability of the tests depend upon correct sampling, accuracy in reading the tests, and an understanding of the limitations of each procedure. Some important variables include such things as the timing of the sampling period (before or after brushing and flossing), early or late in the day (food accumulation), and the nature of the diet (high in sucrose). Also, variation among individuals, especially in their susceptibility to dental caries, often influences test results. Because of these potential variables, it is recommended that saliva be collected before breakfast and before brushing one's teeth. For accuracy, it should be repeated several times.

LABORATORY OBJECTIVES
This exercise will use the Regosa test to determine susceptibility of individuals to dental caries. At the completion of this exercise, you will
- understand the relationship between fermentation and the production of acid in saliva and on the teeth;
- understand the procedure used in this exercise to estimate bacterial activity;
- understand the relationship between dietary sugar and the activity of streptococci and lactobacilli in the oral cavity.

MATERIALS NEEDED FOR THIS LAB
1. One tube of Snyder test agar for each saliva sample being tested.
2. Sterile test tube.
3. Sterile pipettes.

LABORATORY PROCEDURES
1. Obtain one tube of Snyder test agar. Melt the tube and hold at 50°C.
2. Chew the paraffin block to stimulate production of saliva. Collect at least 2.0 ml of saliva in a sterile test tube.
3. Transfer 0.2 ml of saliva into the tube of Snyder test agar. Mix thoroughly and allow to solidify as an agar tall.
4. Incubate at 37°C.
5. Record the color of the Snyder test at 24-hour intervals.
6. Record your results on the Laboratory Report Form.

Name _____

Section _____

LABORATORY REPORT FORM

EXERCISE 35
LACTOBACILLUS ACTIVITY IN SALIVA

What is the purpose of this exercise?

1. Complete the following information for your saliva sample. Use Table 35.1 to assist in your interpretation of the results.

BACTERIAL COUNT FOR SALIVA AND TIME OF COLOR CHANGE		
	RESULTS	INTERPRETATION
Color of Snyder test agar in 24 hours		
Color of Snyder test agar in 48 hours		
Color of Snyder test agar in 72 hours		

2. Complete the following chart, using data from other groups in your laboratory section.

Student Number	Time of Color Change (in hours)	Approximate Number of Bacteria/ml	Number of Fillings or Cavities
1			
2			
3			
4			
5			

QUESTIONS

1. What correlation, if any, can be observed in the data collected and displayed above?

2. List at least three factors that might affect the results of these tests.

3. How would you expect the results to change after the use of mouthwash?

4. How would you expect the results to change after eating candy?

5. How would you explain results showing that one of the students in your sample had a very high susceptibility to dental caries on the basis of the Snyder test, yet actually had a very low number of dental caries and/or fillings?

VIRAL POPULATION COUNTS:
PLAQUE COUNTING AND PLAQUE MORPHOLOGY

This exercise involves the plaque formation by viruses on permissive hosts. We will be both counting the number of plaques formed and determining the specific plaque morphology.

BACKGROUND

A **bacteriophage** is a virus that parasitizes bacteria and, like most viruses, lyses the cell that serves as its host. If the cells that the viruses lyse are immobilized in soft agar, a clear area, or **plaque**, is produced. The plaque appears as a circular, clear zone in an otherwise homogeneously turbid or opaque field. The cleared area results, of course, from the lysis of the bacterial cells in (or on) the agar.

Viruses that infect animal or human cells also destroy their host cells, producing plaques when cultured in tissue-culture systems that use tissue monolayer growing on agar surfaces. However, not all animal cells grow well on agar layers; some must be cultured in rotating glass tubes with the cells growing on the inner surface of the glass, which is constantly bathed with medium. When these tissue-culture systems are used, distinct plaques do not develop. Instead, a characteristic change in cell morphology and physiological activity precedes outright cell death. These changes are often referred to as **cytopathic effect (CPE)**.

Plaques are analogous to colonies of bacteria growing on a nutrient medium. The bacteria in the agar are the medium on which the viruses grow. Like a bacterial colony, a plaque can be assumed to represent the progeny of a single bacteriophage. The number of plaques can be used to count the number of viruses in a suspension. The titration of a bacteriophage suspension is simply the determination of the number of plaque-forming units (PFU) present in 1 ml. The most frequently used method for counting bacteriophage is to prepare serial dilutions and then to plate the dilutions on lawns of bacterial cells.

Bacteriohage plaques show a typical plaque morphology (remember colonial morphology?). When plated on suitable permissive host cells, the morphology of the plaque is characteristic of the bacteriophage and the strain of host cells they were grown on. Of course, if the bacterial strain is not a suitable host (nonpermissive) for the virus, plaques will not be observed at all. While the most apparent trait is the diameter of the plaque, other distinctive characteristics can also be observed. These may include:

- Sharpness of the plaque edge
- Presence of a halo
- Relative size of the halo
- Clarity of the plaque.

PLAQUE MORPHOLOGY OF T4 AND T4r

Bacteriophage T4r is a mutant strain of T4 that produces a dramatically different type of plaque. The *r* stands for **rapid lysis**—bacteriophages with this genetic characteristic (the ability to cause rapid cell lysis) produce sharp plaque borders with centers that are clear and free of bacteria. In contrast, T4 will produce a plaque with hazy edges, resembling a halo of reduced turbidity around a partially cleared center. The host strain of bacteria for the T4 family of phages is *Escherichia coli* B.

EXPERIMENTAL DESIGN

The standard plate count, as it is used for counting bacteria (see Exercise 22), cannot be used to count viruses. Reproducible plaque counts and consistent plaque morphology require that the bacteria and viruses be suspended together in the agar during the time the plaques are developing. The most convenient way of doing this is to employ the agar overlay technique for the actual plaque count.

In the **agar overlay technique**, the virus suspension is mixed with the bacteria in about 3 to 5 ml of melted agar. The agar is gently mixed and then poured over an agar layer that has already solidified in petri plates. The lower layer of agar provides a nutrient base for the bacterial and viral growth occurring in the upper layer. Restricting plaque formation to a relatively thin surface layer ensures vigorous virus and bacterial growth and clear, easily seen plaques.

In this exercise you will titrate at least one virus suspension. Your instructor may have you do more than one titration so that you can observe different types of virus plaque morphologies, or varied viral strains may be assigned to alternating students. By varying the strain of bacteriophage, differences in the shape of the plaque can easily be observed. Study the flow chart shown in Figure 36.1 and refer to it as you proceed. If too many bacteriophages are plated, crowded or confluent plaques will be produced and it will be difficult, perhaps impossible, to accurately study their morphology.

LABORATORY OBJECTIVES

Viruses are obligate parasites that must be grown on living cells. A virus culture is, in fact, a culture of two organisms—a virus and its host cell. In this exercise you should

- understand the agar overlay technique and the use of serial dilutions for the titration of a virus suspension;
- understand what a plaque is and why it is analogous to a virus colony;
- learn to distinguish different viruses by their plaque morphology;
- understand how viruses cause the destruction of their host cell and appreciate why virus growth in a living organism is usually a serious pathogenic condition;
- understand the nature of rapid lysis, as the term is used in this exercise;
- realize the biological consequences of plaque formation for the host cells and understand the clinical significance of host-cell destruction *in vivo*.

MATERIALS FOR THIS LAB

The following materials are needed for the titration of a single bacteriophage:

1. Five prepoured TSA petri plates. The plates should be warmed in the incubator prior to use.
2. Five tubes of 3 ml soft TSA (0.7% agar), melted and kept at 50°C.
3. Six tubes of 9.0 ml TS broth.
4. 18–24 hour culture of *Escherichia coli* B.
5. Suspensions of bacteriophage (T4 or T4r as assigned by instructor). About 2 ml will be needed and should have a titer of about 10,000 bacteriophages/ml.
6. Sterile pipettes.

LABORATORY PROCEDURE

1. Be sure to refer to Figure 36.1 as you proceed.
2. Obtain and label five petri plates and five tubes of soft plating agar. The plates should be labeled with the dilution (1:10 to 1:100,000) and the bacteriophage type and host-cell strain. You will also need six dilution blanks containing 9.0 ml of TS broth.
3. Maintain the soft agar in a 50°C water bath to prevent it from solidifying.
4. Transfer about 5 ml of TS broth from one of the tubes to the slant of *Escherichia coli* B. Rock the slant back and forth to suspend the bacterial cells. Label the remaining five blanks according to the serial dilution series (1:10 to 1:100,000).
5. Prepare a serial dilution of the bacteriophage suspension. Transfer 1 ml from the undiluted stock suspension to the first dilution blank. Mix well, then transfer 1 ml to the next blank. Repeat these transfers until all dilutions have been made.
6. Transfer two drops of the bacterial suspension into each of the soft agar tubes.
7. Transfer 1 ml from the first bacteriophage dilution (1:10) into the first soft agar tube. Mix gently and immediately pour into the appropriately labeled plate.
8. Repeat this procedure until all the dilutions have been plated.
9. Incubate all plates at 37°C for 24 hours. If you cannot complete the exercise after 24-hour incubation, store the plates in the refrigerator until the next laboratory period.
10. Count the plaques on all dilutions with 25 to 250 plaques present. Observe the morphology of the plaques observed (see Figure 36.2)
11. Complete the Laboratory Report Form for this exercise.

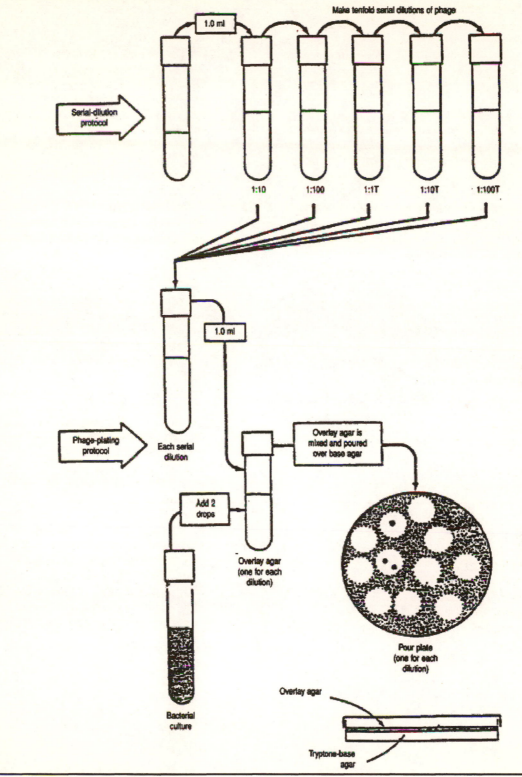

Figure 36.1   Plaque count procedure.

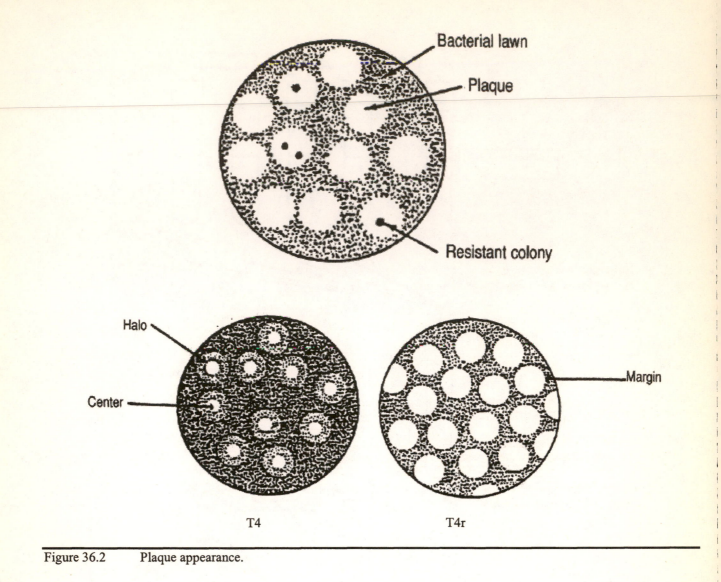

Figure 36.2   Plaque appearance.

LABORATORY REPORT FORM
EXERCISE 36
VIRAL POPULATION COUNTS

What is the purpose of this exercise?

1. Complete the following table. You should obtain any necessary data from other students in the class and use their plates to determine plaque size and morphology.

Bacteriophage (Host)	RESULTS OF PLAQUE ASSAY			
	Count (PFU/ml)	Size of Center (mm)	Size of Halo (mm)	Morphology and Margin
T4 (*E. coli* B)				
T4r (*E. coli* B)				

> **NOTE:** The plaque counts must be calculated for the number of plaques that formed on the plates. You should determine the number of plaque-forming units in each milliliter of stock suspension (PFU/ml).

2. Did you observe any variations in the plaque morphology of any of these bacteriophages? If so, what is the significance of such variation?

3. How does your plaque count compare with the expected count, based on the reported titers of the bacteriophage suspension? Explain any differences.

4. Did any of the plaques have colonies growing within the cleared area? If so, what is the significance of those colonies?

QUESTIONS

1. Describe the agar-overlay technique.

2. What is rapid lysis?

3. Compare the number of plaques on each of your dilutions. Do they agree with the dilution factor? (Does the 1:10 plate have 10 times as many plaques as the 1:100 plate?)

FOOD MICROBIOLOGY

Most foods provide an excellent growth medium for bacteria and other microorganisms. The supply of organic matter is plentiful, the water content is usually sufficient, and the pH is neutral or only slightly acidic. The result is food spoilage, which implies an economic loss to the manufacturer and a waste of money to the consumer, as well as a threat to health. Many types of food spoilage are aesthetically displeasing, and we refrain from eating the food even though the spoilage does not pose a life-threatening situation. In this exercise we will examine the presence of bacteria on various food products.

BACKGROUND

Microorganisms can be introduced into food at any time along its route from farm to our table. Plants will contain organisms found in soil, water, and air naturally occurring on their surfaces. In addition, organisms may penetrate the surface and be present in deeper tissue. Animals also are going to contain significant numbers of organisms as part of their normal flora. During the slaughtering process, these organisms may be introduced to the muscle tissue— an area normally free of microorganisms. These food products are then processed in some manner and stored for future use. Here it is possible for equipment and handlers to once again introduce organisms into the food product. Many, if not most, of these organisms are destroyed in the final processing of these products (canning, freezing, pasteurizing, drying, etc.). Even so, some organisms may remain, and more will likely be introduced as this product is readied for eating where it again encounters utensils and handlers.

It is these organisms which remain in our food products that can contribute to food spoilage and food-borne illness. To minimize potential food-borne problems, food-processing companies and public health agencies perform standard plate counts on food samples. These determine the overall quality of the food, the cause of spoilage, the existence of proper storage methods, and the potential presence of pathogens.

The standard plate count involves diluting small samples of food, plating it using a general media, incubating, and determining the number of organisms present. Whereas these results are indicative of overall food quality, the information gained is limited. If varied incubation temperatures are used, it is also possible to determine the relative presence of thermophiles or psychrophiles. (Why would this be important?) In addition, various selective and differential media can be used for the detection of specific groups of organisms, such as coliforms (MacConkey agar), *Salmonella* or *Shigella* (Hektoen enteric or SS agar), and *Staphylococcus aureus* (Mannitol-salt agar). Not only is this information important in determining the overall quality of food products, but it also aids in the development of adequate preservation methods for varied products and in assessing their effectiveness.

Bacterial counts of foods are not, however, indicative of food safety. It is possible for a product to have relatively high total counts and yet be completely safe to eat. For this reason, total counts are of little value when determining the quality of fermented food products (e.g., yogurt or pickles). Plate counts also do not indicate the presence of microbial toxins in foods. Consequently, it is important that we remember that the total plate count of a food product is only an indication of the organisms that are present and that can grow under the incubation conditions used.

LABORATORY OBJECTIVES

In this exercise you will be performing plate counts to determine the quantity of bacteria present in/on varied food products. At its completion you should be able to

1. perform and evaluate a quantitative analysis of the bacteria in food samples, and
2. understand the significance of bacterial plate counts when applied to food.

MATERIALS NEEDED FOR THIS LAB

The following will be needed for each food sample being tested:

1. Plate count agar talls (5). These will need to be melted and held at 50°C to keep them liquid.
2. Sterile 9.0 ml water blank OR sterile 90.0 ml water blank (1).
3. Sterile 99.0 ml water blank (2).
4. Sterile pipettes, 1.0 ml and 10.0 ml.
5. Sterile petri plates (5).
6. Balance.

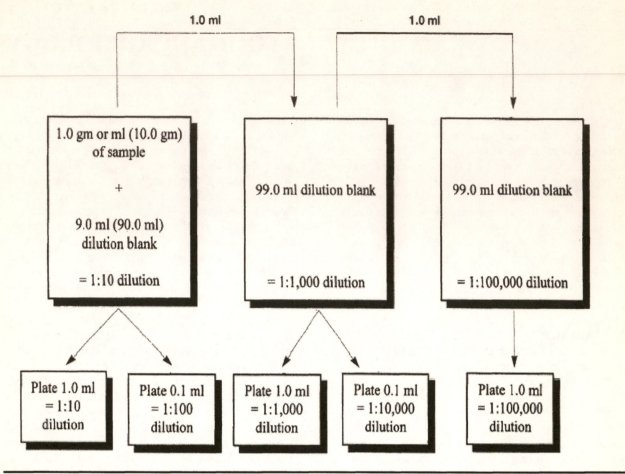

Figure 37.1   Food-testing procedure.

---

7.   Sterile blender or mortar and pestle

LABORATORY PROCEDURE

Refer to the flow chart in Figure 37.1 as you perform this exercise.

1.   Obtain five sterile petri plates and label them 1:10, 1:100, 1:1,000, 1:10,000, and 1:100,000.
2.   Obtain one sterile 9.0–ml water blank and two sterile 99.0–ml water blanks.
3.   Aseptically weigh 1 g or pipette 1 ml of the food to be tested.
      a.   If the food is solid, it should be placed into the blender or mortar. Gradually add the 9.0–ml sterile water blank, mixing the food and fluid thoroughly with the blender or the grinding of the pestle.
      b.   If the food sample is liquid, directly add it to the 9.0–ml sterile water blank and mix. This will give you a 1:10 dilution.

NOTE: If 1 g of the food sample will not provide a representative sample of the product, you may weigh out 10 g of the food and add it to 90–ml of sterile water. This will also provide a 1:10 dilution.

4.   Pipette 1.0 ml of your dilution into the petri plate labeled 1:10 and 0.1 ml of your dilution into the petri plated labeled 1:100. Be careful to avoid airborne contamination.
5.   Pipette 1.0 ml of the dilution to the 99.0–ml sterile water blank. Mix thoroughly. Pipette 1.0 ml of the dilution into the petri plate labeled 1:1,000 and 0.1 ml into the plate labeled 1:10,000.
6.   Pipette 1.0 of the 1:1,000 dilution into the second 99.0 ml sterile water blank. Mix thoroughly. Pipette 1.0 ml of the dilution to the petri plate labeled 1:100,000.
7.   Obtain five tubes of melted plate count agar that has been maintained at 50°C.

8. Aseptically pour a melted plate count agar tall into each petri plate, rotating each carefully to mix the agar and dilution. Allow the media to harden.
9. Incubate at 37°C for 24 to 48 hours.
10. Count the colonies, both surface and subsurface, in your plates. Record the results on the Laboratory Report Form. Use the term "TNTC" (too numerous to count) to refer to those plates with well over 300 colonies. Select the dilution with a final count of between 30 and 300 colonies and calculate the total plate count per gram (ml) of food.

NOTES _____

Name _____

Section _____

LABORATORY REPORT FORM

EXERCISE 37
FOOD MICROBIOLOGY

What is the purpose of this exercise?

The food tested was _____

Plate–count results:

Dilution	1:10	1:100	1:1,000	1:10,000	1:100,000
Plate Count					

Calculate the cfu/g or ml of your food sample.

Dilution used _____

Count results _____

cfu/g or ml _____

Record the results for the other food products tested by your class.

FOOD SAMPLE	cfu/g or ml

QUESTIONS

1. Does a high standard plate count in food indicate that it must not be eaten? Why or why not?

2. What errors may account for inaccurate plate count results?

3. You have gotten very similar total counts (approximately 10^7 cfu/g) for both ground beef and potato salad from the deli. How would you interpret these results as to the safety of food consumption?

4. You have a can of soup with bloated ends. You know that you should not eat it as the food is very likely spoiled and decide that you want to see what microorganism(s) is(are) present. What incubation conditions would you use when testing this product? Be sure to include temperature, oxygen, and pH of media for incubation.

5. When testing for bacteria in shell fish, MacConkey agar is often used instead of plate count agar. Why would a selective/differential media be chosen? What type of organism is being tested for and why?

WATER ANALYSIS: STANDARD METHODS

Freshwater supplies are often subjected to the introduction of raw sewage, inorganic and organic industrial wastes and fecal contamination by all animals, including humans. This can result in the occurrence of dead fish on the beach, or an outbreak of typhoid fever or cryptosporidium in the human population. In this exercise we will be examining the presence of bacteria in standing bodies of water through the utilization of a standard method analysis. In Exercise 39, we will examine the application of the membrane filter technique.

BACKGROUND

A multitude of pathogenic organisms can be transmitted through the ingestion of contaminated water. These include *Salmonella* and *Shigella* sp., gastroenteritis and hepatitis viruses, and the protozoa *Giardia, Entamoeba,* and *Cryptosporidium.* To test water for each potential pathogen is not practical. Instead, our water safety standards have been based on the presence (or better, the absence) of readily determined, usually nonpathogenic fecal organisms. If a water supply is found to contain any microorganisms of fecal origin, it might also contain pathogens and, therefore, should not be considered safe to consume. Those bacterial organisms that are commonly used as indicators of fecal contamination are referred to as the *coliform bacteria*.

COLIFORMS

Coliforms are those bacteria that have characteristics similar to *Escherichia coli.* They are aerobic or facultative, gram-negative, nonspore-forming bacilli that ferment lactose with the production of acid and gas. In addition, they will grow on EMB (eosin–methylene blue) agar. These organisms are usually nonpathogenic; however, they are usually found in large numbers in animal feces (as many as 10^6 to 10^9 cells/g of feces) and survive longer in cold surface water than do most intestinal pathogens. In addition, direct correlations have been shown between the presence of these indicator organisms and the degree of fecal contamination. Another extremely significant reason to utilize these organisms is their relative ease of detection.

The coliform group of bacteria includes (in addition to the *Escherichia*) *Enterobacter, Citro–bacter,* and *Klebsiella* organisms. These other organisms, however, may also enter the water via soil and vegetation runoff and so are not strictly enteric in origin. Because of this, some prefer to screen for the presence of fecal coliforms, or those organisms that are gram-negative, nonspore-forming bacilli that ferment lactose producing acid and gas at 44.5°C in 24 hours. By using an elevated temperature of incubation, many of the nonenteric coliforms are eliminated.

Recent studies have indicated that the fecal enterococci might serve as more accurate indicator organisms. These gram-positive enteric organisms are less likely to be introduced by nonfecal methods but are not as readily tested for.

BACTERIOLOGICAL EXAMINATION OF WATER

Two methods have routinely been used for the determination of water contamination. The first is the plate count method, which involves the plating of dilutions of water on nutrient agar and incubating at 30°C and 35°C. It is well understood that this will not indicate the total number of organisms in the water, but only those that can grow under the incubation conditions used. It can, however, indicate the relative number of organisms present in water. This method does not distinguish between environmental and fecal microorganisms so it is not commonly used for determining of bacterial quality of water supplies.

The second major method involves the utilization of indicator organisms, such as the coliforms, which are present in large numbers in fecal material. It should be mentioned at this time that indicator organisms are not always effective in showing the potential presence of pathogens. More sophisticated testing methods give reason to suspect that enteric pathogens may be present in the absence of indicator organisms due to their increased resistance to the aquatic environment as well as to chlorination of water. Despite these limitations, indicator organism screening continues to be the most common method of bacteriologic analysis of water. Commonly performed testing protocols include the Standard Methods Analysis and the membrane filter technique. Both of the methods mentioned, however, are somewhat limited in their ability to determine the level of coliform contamination.

MOST PROBABLE NUMBER (MPN) ANALYSIS

The most probable number analysis, or MPN, while not commonly used in clinical applications, is extensively used in public health and environmental microbiology. There are two important advantages of the MPN procedure. First, it is a good screening procedure, useful when a high level of precision is to required. (Do you need to know *exactly* how many bacteria are present, or is it sufficient to know *about* how many are present?) The second advantage is that selective and differential media can be used. This makes it possible to estimate numbers of specific kinds of bacteria, such as coliforms in water. It must be noted, however, that the data obtained from an MPN test is only an estimate of the most *probable* number of bacteria in a given volume of sample.

The MPN test is accomplished by inoculating three sets of tubes with known volumes of sample. Typically, there are three or five tubes of media in each of three sets. The first set usually contains 10.0 ml of double-strength broth. Each tube in this set is 10.0 ml of sample. Double-strength broth is used so the medium will not be too dilute when you add an equal volume of sample. The second and third sets of tubes contain 10.0 ml of single-strength broth and are inoculated with 1.0 ml and 0.1 ml of sample, respectively (see Table 38.1)

Table 38.1   Most Probable Number Sets

Set	Contains 10 ml of	Volume of Sample Used
First	Double-strength broth	10.0 ml
Second	Single-strength broth	1.0 ml
Third	Single-strength broth	0.1 ml

After a suitable incubation period, the pattern of growth obtained is noted and compared to a table of statistically determined most probable numbers (see Table 38.2). This table is based on five tubes in each set. If there are 5 tubes in a set and 3 of the 10.0–ml tubes and 1 of the 1.0 ml tubes showed growth (+++---/+-----/------), the most probable number would be 11 bacteria per 100 ml of sample. The MPN procedure estimates the most probable number of bacteria likely to be present in 100 ml of the sample. The MPN tables give that value for each of the possible combinations of growth in the three sets of tubes. In addition to the probable numbers present, the table gives the 95% confidence limits for the given test result. This tells us that, for the test example given above, there is 95% assuredness that

the actual value is between 4.0 and 29 organisms/100 ml of sample.

STANDARD METHODS ANALYSIS

The Standard Methods Analysis of water is a three-step process—the presumptive test, the confirmed test, and the completed test.

1. **Presumptive Test.** The presumptive test is an enrichment procedure for the presence of coliforms using a selective medium. Lauryl sulfate tryptose lactose (LST) broth is inoculated with aliquots of the water and incubated. This medium contains lauryl sulfate, a surface active detergent that inhibits the growth of gram-positive organisms, and lactose. The presence of acid and gas in 24 to 48 hours is considered presumptive evidence for the presence of coliform organisms. When the presumptive test is performed in this manner it is simply a qualitative test to answer the question, "Are there any coliform organisms present?" Whereas this is important to know, it is even more significant to know the degree of coliform contamination. The presumptive test can be used as a MPN determination by using multiple tubes of LST.

2. **Confirmed Test.** The presence of a positive presumptive test is confirmed with brilliant green lactose bile (BGLB) broth. It is inoculated from a positive presumptive tube and incubated at 35°C. This medium further inhibits the presence of any gram-positive organisms by the introduction of bile salts and brilliant green. The production of gas within 24 to 48 hours "confirms" the presence of coliform organisms.

    A variation of the confirmed test that is often performed involves the additional inoculation of FC medium, which is then incubated at 44.5°C. The FC broth is also a lactose and bile salts medium that further inhibits nonfecal organisms with its elevated temperature of inoculation It is, therefore, highly selective for the fecal coliforms.

3. **Completed Test.** The completed test can be performed to further verify the presence of *Escherichia coli* and to distinguish its presence from that of *Enterobacter aerogenes* (commonly found in soil and plant runoff). It may involve several testing procedures including
    a. EMB agar for presence of characteristic growth,
    b. TSA slant for Gram-staining,
    c. LST or BGLB broth for confirmation of lactose fermentation, and

d. IMViC testing for presence of typical test patterns (see Exercise 34 for further discussion of the IMViC tests).

When the presumptive and confirmed tests show distinctive positive results the completed test is not considered essential for the confirmation of the presence of coliform organisms and so is not performed.

LABORATORY OBJECTIVES

In this exercise you will be performing the Standard Methods Analysis on a water sample. Through this exercise you should

- understand the Standard Method Analysis of water and the three steps involved in the complete analysis;
- know the basis for selective and differential media used in water analysis;
- be able to interpret the results obtained as to overall water quality.

MATERIALS NEEDED FOR THIS LAB

If desired, your instructor may have you bring in a water sample from home, a nearby stream, a lake or some other location for testing. You should use a sterile collection bottle (available from your instructor). Surface water (lake, pond, stream, etc.) samples should be taken under the surface so as to prevent the introduction of material floating on the surface. If the sample cannot be collected within 1 to 2 hours of testing, keep it refrigerated until your laboratory period.

1. Water sample.
2. Five tubes of lauryl sulfate tryptose lactose (LST) broth (double strength), 10 ml/tube, containing Durham fermentation tube.
3. Ten tubes of lauryl sulfate tryptose lactose (LST) broth (single strength), 10 ml/tube, containing Durham fermentation tube.
4. One tube of brilliant green lactose bile (BGLB) broth containing Durham fermentation tube.
5. One plate eosin–methylene blue (EMB) agar.
6. Pipettes, 10.0 ml and 1.0 ml.
7. Pipetter

LABORATORY PROCEDURE

Before starting this procedure, review the protocol as shown in Figure 38.1.

1. Obtain and label five tubes of double-strength LST broth. Inoculate each with 10.0 ml of your water sample. Be sure to carefully mix the water sample prior to sampling.
2. Obtain and label 10 tubes of single-strength LST broth. Inoculate five tubes with 1.0 ml of your water sample in each. Inoculate the remaining five with 0.1 ml of water sample in each.
3. Incubate all tubes at 37°C for 24 to 48 hours.
4. Following incubation, determine the presence of acid and gas in the tubes. Record your results on the Laboratory Report Form. Using Table 38.2, determine the MPN value and confidence levels for your sample.
5. From one of your most dilute positive LST broth tubes, inoculate a tube of BGLB broth. If your water sample was negative at all dilutions, use a positive presumptive test tube from another student to complete the testing procedure.
6. Incubate the tube at 37°C for 24 hours. If you are unable to examine the tube and complete the testing at that time, refrigerate the tube until your next laboratory period.
7. Following incubation, read the results of the BGLB broth. Record the results on the Laboratory Report Form.
8. From a positive BGLB test, streak a plate of EMB agar.
9. Incubate the plate at 37°C until your next laboratory period.
10. Following incubation, examine the EMB plate for the presence of colonies with a green "metallic" sheen. These colonies are positive for the fermentation of lactose. Record the results on the Laboratory Report Form.
11. Perform a Gram-stain on a representative colony.

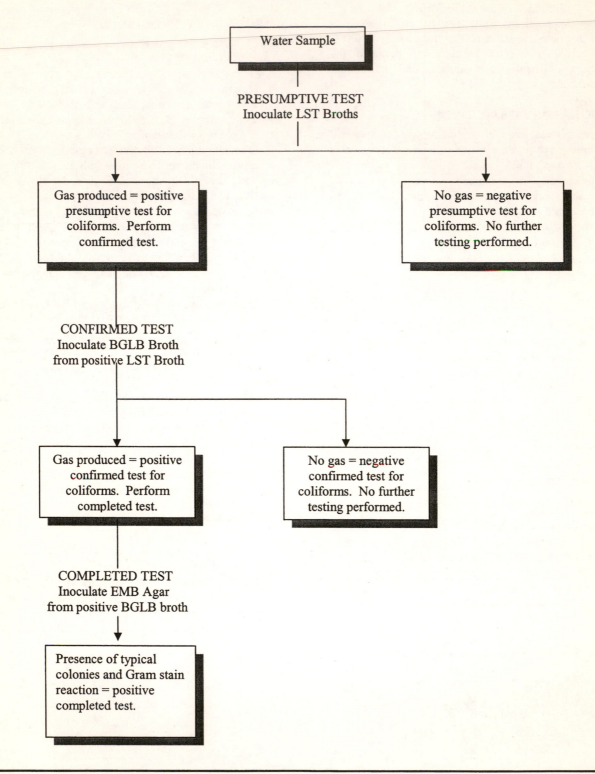

Figure 38.1   Standard Methods Analysis of water protocol testing.

Table 38.2   The MPN Index and 95% Confidence Limits for Various Combinations of Positive Results When Five Tubes Are Used per Dilution

Number of Positive Tubes			MPN/ 100 ml	95% Confidence		Number of Positive Tubes			MPN/ 100 ml	95% Confidence	
10.0 ml	1.0 ml	0.1 ml		Lower	Upper	10.0 ml	1.0 ml	0.1 ml		Lower	Upper
0	0	0	<2	--	--	4	3	0	27	12	67
0	0	1	3	1.0	10	4	3	1	33	15	77
0	1	0	3	1.0	10	4	4	0	34	16	80
0	2	0	4	1.0	13	5	0	0	23	9.0	86
1	0	0	2	1.0	11	5	0	1	30	10	110
1	0	1	4	1.0	15	5	0	2	40	20	140
1	1	0	4	1.0	15	5	1	0	30	10	120
1	1	1	6	2.0	18	5	1	1	50	10	150
1	2	0	6	2.0	18	5	1	2	60	30	180
2	0	0	4	1.0	17	5	2	0	50	20	170
2	0	1	7	2.0	20	5	2	1	70	30	210
2	1	0	7	2.0	21	5	2	2	90	40	250
2	1	1	9	3.0	24	5	3	0	80	30	250
2	2	0	9	3.0	25	5	3	1	110	40	300
2	3	0	12	5.0	29	5	3	2	140	60	360
3	0	0	8	3.0	24	5	3	3	170	80	410
3	0	1	11	4.0	29	5	4	0	130	50	390
3	1	0	11	4.0	29	5	4	1	170	70	480
3	1	1	14	6.0	35	5	4	2	220	100	580
3	2	0	14	6.0	35	5	4	3	280	120	690
3	2	2	17	7.0	40	5	4	4	350	160	820
4	0	0	13	5.0	38	5	5	0	240	100	940
4	0	1	17	7.0	45	5	5	1	300	100	1300
4	1	0	17	7.0	46	5	5	2	500	200	2000
4	1	1	21	9.0	55	5	5	3	900	300	2900
4	1	2	26	12	63	5	5	4	1600	600	5300
4	2	0	22	9.0	56	5	5	5	>1600	–	–
4	2	1	26	12	65						

(From: *Standard Methods for the Examination of Water and Wastewater*)

NOTES _____

LABORATORY REPORT FORM

EXERCISE 38
WATER ANALYSIS: STANDARD METHODS

What is the purpose of this exercise?

Source of water tested _____

PRESUMPTIVE TEST:

1. Record the number of tubes at each dilution that were positive for the presence of both acid and gas following incubation.

    10.0 ml   _____

    1.0 ml   _____

    0.1 ml   _____

2. Using Table 38.2, determine the MPN for your water sample.

    MPN   _____

    95% confidence limits   _____

CONFIRMED TEST:

1. Describe the results obtained in the BGLB broth.

2. Was the confirmed test positive or negative?

COMPLETED TEST:

1. Describe the appearance of colonies on the EMB agar.

2.   Gram-stain results:

Morphology   _____

Gram reaction   _____

QUESTIONS

1.   Why are laboratory tests for bacterial pollution of water based on the identification of coliform bacteria and especially *Escherichia coli?*

2.   What could be the cause of a positive presumptive test that does not test positive with the confirmed test?

3.   Could a sugar other than lactose be used in the media for coliform detection? Why or why not?

4.   Can water with a negative presumptive test following 24-hour incubation be considered completely safe to drink?

WATER ANALYSIS:
MEMBRANE FILTER TECHNIQUE

In Exercise 38, the Standard Methods Analysis was utilized to determine the presence of coliform contamination of water. This method works quite effectively when significant numbers of bacteria are present in water. When few organisms are suspected, the Standard Method does not work because 10.0 ml is the largest volume of water that can be used for a single inoculum. When larger volumes need to be tested, the membrane filter technique is the method of choice.

BACKGROUND

The membrane filter technique is widely used for the detection of coliforms in water samples. This method is recommended by the American Public Health Association for the testing of water because of the distinct advantages it has over Standard Methods. These advantages include:

1. direct counts of either coliform colonies or colonies of other organisms can be made. For example, a direct count of *Salmonella* present in water could be more readily determined;
2. the results can be obtained in a shorter time;
3. larger volumes of water can be tested; and
4. the results are more accurate and are reproduced more readily when appropriate media and incubation conditions are used. In addition, this testing protocol is less expensive to perform.

Volumes of water are filtered through a membrane filter with a pore size of 0.45 μm. The filter can then be directly placed on either an agar medium or a blotter pad saturated with broth. For example, the filter can be placed on a pad saturated with m-Endo MF broth and incubated at 35°C for detection of total coliform counts. If the filter is placed on m-FC broth and incubated at 44.5°C, fecal coliform determinations can be made; incubation of KF streptococcus agar at 35°C will give fecal streptococci values. The use of membrane filters, however, is not without its limitations. To determine accurate results, appropriate dilutions of water sample must be made to give final results of 20 to 200 colonies per plate. In addition, if the water is highly turbid, contains silt or large numbers of bacteria or algae, it may clog the filter, making accurate determinations difficult.

LABORATORY OBJECTIVES

- Understand the membrane filter technique of water analysis.
- Compare the utilization of the membrane filter technique with that of the Standard Methods Analysis for the examination of water for bacterial contamination.

MATERIALS NEEDED FOR THIS LAB

1. Water sample.
2. Media, which may include:
     m-Endo MF broth
     m-FC broth
     KF Streptococcus agar.
3. Bacteriological membrane-filter assembly, sterile.
4. Sterile membrane filters, 0.45-μm pore size.
5. 47 mm petri plate, sterile.
6. Absorbent pads, sterile.
7. Sterile distilled water, 50 ml.
8. Pipette, 5.0 ml.
9. Forceps.
10. 95% ethyl alcohol.

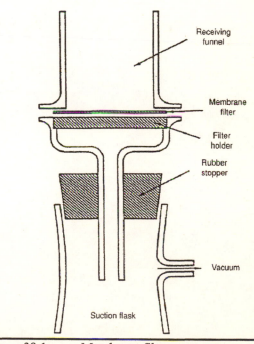

Figure 39.1   Membrane filter apparatus.

LABORATORY PROCEDURE

1. Sterilize a pair of forceps by dipping them in alcohol and then flaming to burn away the alcohol. Aseptically place a sterile absorbent pad in the bottom of each 47-mm petri plate.
2. Pipette 2 ml of m-Endo MF broth onto the absorbent pad. Replace the lid on the plate.
3. Repeat steps 2 and 3 for other media as instructed.
4. Assemble the membrane filter apparatus as shown in Figure 39.1.
5. Aseptically, using sterile forceps, place a sterile membrane on the filter base.
6. Pour 50 ml of sterile distilled water in the receptacle. Pipette 1.0 ml of water sample into the receptacle.
7. Gently turn on the vacuum until the water has been completely drawn through the filter.
8. Using sterile forceps, transfer the filter to the surface of the absorbent pad. Carefully place it on the pad so as to not trap any air under the filter.
9. Incubate the petri plates at 37°C for 24 hours. The plates can be placed in sealable bags to prevent their rapid dehydration.
10. Repeat the filtering process, as directed by your instructor, for plating on additional media. All plates except m-FC broth plates should be incubated at 37°C. The m-FC broth plates should be incubated at 44.5°C.
11. Determine the number of colonies per plate for each media used. Calculate the number of organisms per 100 ml of your water sample. Record your results on the Laboratory Report Form.

LABORATORY REPORT FORM

EXERCISE 39
WATER ANALYSIS: MEMBRANE FILTER TECHNIQUE

What is the purpose of this exercise?

Source of the water tested _____

Media Used	Selective For	Actual Colony Count	cfu/100 ml water
m-Endo Broth MF			
m-FC Broth			
KF Streptococcus Broth			

QUESTIONS

1. What limitations are there to using the membrane filter technique?

2. Explain why the filtration of contaminated water through a membrane filter does not necessarily guarantee that it is safe to drink.

WRITING A LABORATORY REPORT

The accepted style for laboratory reports will be that which is found in standard scientific journal format. This format requires that you use and demonstrate your ability to organize ideas logically, think clearly, and express yourself accurately and concisely. The following sections must be included. Introduce each section of your report with the proper headings as outlined below.

TITLE AND AUTHOR

The title should be descriptive in nature. Remember, this is not a creative writing exercise; it is the explanation of the problem, experimental design, and conclusion of a scientific experiment.

INTRODUCTION

The Introduction, consisting of one or two paragraphs, should

- describe **why** the study was undertaken ("The purpose …");
- present background information about the subject; and
- state the problem to be solved as a hypothesis.

The hypothesis should predict what will happen based on proper use and interpretation of experimental results (not what you **think** will happen). It should be stated as an "If … then …" statement.

MATERIALS AND METHODS

The Materials and Methods section of your report should be written in such a manner as to enable a reader to repeat all tests exactly as you did them and to achieve the same results. It is not necessary to list everything used and the exact steps performed; rather you should direct the reader to the exact source of the test protocols. A few reminders:

- Include any changes made in the stated protocol.
- Do NOT include any unnecessary information (i.e., "the loop was flamed").
- Do NOT simply list Exercise numbers, but indicate which test(s) were performed.

RESULTS

The specific observations made during the experiment should be presented in the Results section. Graphic presentation and/or tables may be included. The results should NOT be included with the Materials and Methods! Remember: The Results section is

- **not** the place to discuss **why** the experiment was performed;
- **not** the place to discuss **how** the experiment was performed;
- **not** the place to give your conclusion or to identify your unknown organism;
- **not** the place to **discuss** whether the results were expected, unexpected, disappointing, or interesting;
- to **simply present the results**, not interpret them.

DISCUSSION AND CONCLUSION

This section should be the major component of your report. Here you will give the Genus and species of your unknown organism and the basis of your decision. Did your data support your hypothesis? What tests helped you determine the identity of your unknown? Which results did not appear to agree with the documented results of your organism? What might you have done differently? Were there any tests that you did not perform that would have aided in the identification of your unknown?

LITERATURE CITED

All references (textbook, laboratory manual, handouts, reference books) that you have used or cited in your report should be listed here. The following format should be followed:

Author. Title. Publication (journal). Publisher: (book). Year:Page(s).

Writing Tips:

1. Write in the past tense, not the present.
2. Avoid the use of personal pronouns (I, we, etc). Instead of writing "I performed …" write that "…was performed."
3. Use metric units and the Celsius temperature scale.
4. Be sure to use proper scientific names for all organisms.
   - Italicize or underline all scientific names;

- capitalize the first letter of genus name and lower case for species;
- when you first mention an organism, you must write the entire genus and species name (i.e., *Escherichia coli*); in subsequent references it may be abbreviated (i.e., *E. coli*).

5. Write concisely and logically. Avoid boring the reader with copious verbiage and excessively formal writing!
6. Your report should be written in grammatically correct English.
7. Reports should be typed and double-spaced.

ALGAL CULTURES
Gleocapsa

BACTERIAL CULTURES
Acinetobacter calcoaceticus
Bacillus cereus
Bacillus megaterium
Bacillus stearothermophilus
Bacillus subtilis
Clostridium sporogenes
Corynebacterium diphtheriae
Corynebacterium xerosis
Enterobacter aerogenes
Enterococcus faecalis
Escherichia coli
Halobacterium salinarium
Klebsiella pneumoniae
Lactobacillus sp.
Lactobacillus delbruckii subspecies *bulgaricus*
Lactococcus lactis
Micrococcus luteus
Moraxella (Branhamella) catarrhalis
Mycobacterium smegmatis
Neisseria gonorrhoeae
Proteus mirabilis
Proteus vulgaris
Pseudomonas aeruginosa
Pseudomonas fluorescens
Salmonella typhimurium
Serratia marcescens
Shigella flexneri
Staphylococcus aureus
Staphylococcus epidermidis
Streptococcus pneumoniae
Streptococcus pyogenes
Vibrio parahaemolyticus

MOLD CULTURES
Aspergillus
Penicillium
Rhizopus

BACTERIOPHAGE CULTURES
Coliphage T4
Coliphage T4r
Host Bacteria: *E. coli* B

PROTOZOA CULTURES
Amoeba
Euglena
Paramecium

YEAST CULTURES
Saccharomyces cerevisiae

SOURCES
Smaller laboratories may find that commercially prepared cultures are often more convenient than keeping stock cultures. This is particularly true of the reference cultures available in dried disk form. Most biological supply companies today are reliable sources of cultures. The bacteriophage demonstration kits produced by Carolina Biological are highly recommended.

Although some programs may find it more convenient to use commercial cultures, quality control is, and always should be, the responsibility of the user.

USE OF PATHOGENS
There has been much discussion in recent years as to the advisability of using pathogenic or opportunistic organisms in the introductory microbiology laboratory. As much as it would be desirable to never use pathogens with first-semester students, there are certain test reactions (i.e., coagulase) that cannot be suitably illustrated without the use of pathogens. For this reason, their use has been minimized, but they continue to be used when necessary.

It is the belief of the author that the most important lesson that can be left with all introductory microbiology students is that all organisms be treated as potential pathogens. With the careful emphasis on proper aseptic technique, the use of these organisms should not put the students at risk.

MEDIA, STAINS, REAGENTS, AND SUPPLIES

MEDIA

The following media are used for the exercises contained in this manual. Most of them are available in dried as well as pre-poured form. In most cases, the expense of the pre-poured form is not justified, although convenience and quality may be overriding considerations in some circumstances.

Alkaline peptone medium
Bile-esculin agar
Blood agar base with 5% sheep blood
Brilliant green lactose bile (BGLB) broth
Chocolate agar, with supplement
Columbia CNA agar
DNase agar with methyl green
EMB agar
Hektoen enteric agar
KF streptococcus broth
Lauryl sulfate tryptose broth
Levine EMB agar
Litmus milk
MacConkey's agar
Mannitol salt agar
m-Endo MF broth
m-FC broth
Minimal salts with glucose agar
Motility medium
MR-VP broth
Mueller-Hinton agar
Mueller tellurite agar
Nitrate broth
Nutrient agar
Nutrient broth
Nutrient broth supplemented with 0.1% yeast extract
       and 1.0% glucose
Nutrient broth with pH 3.0, 5.0, 9.0
Nutrient gelatin
Peptone iron agar
Phenol red broth base with
       Glucose
       Lactose
       Mannitol
       Sucrose
Phenylethanol agar
Plate count agar
Sabouraud agar
Salmonella-Shigella (SS) agar
Selenite F broth

Simmons' citrate agar
Snyder test agar
Starch agar
Sulfur Indole Motility (SIM) agar
Thayer-Martin agar
Thioglycollate medium, fluid
Thiosulfate citrate bile sucrose (TCBS) agar
Triple sugar iron (TSI) agar
Tryptone broth
Tryptic soy agar (TSA)
Tryptic soy agar + sodium chloride (5%, 10%, 15%)
Tryptic soy agar + sucrose (10%, 25%, 50%)
Tryptic soy broth
Tryptic soy soft agar (0.7% agar)
Urea broth

STAINS AND REAGENTS

Acetone/ethanol
Acid alcohol
Albert's stain
Alpha-naphthol, 6%
Barritt's Reagent A
Barritt's Reagent B
Coagulase plasma
Copper sulfate
Crystal violet
Crystal violet, 1% w/v aqueous
Dorner's nigrosin
Ferric chloride, 10%
Gram's crystal violet (Hucker's)
Gram's iodine
Gram's safranin
Gray's solution "A"
Hydrogen peroxide, 3%
India ink
Iodine reagent (Starch hydrolysis)
Kinyoun's carbolfuchsin
Kovac's reagent
Loeffler's methylene blue
Malachite green
Methyl red indicator
Methylene blue
Methylcellulose
Nitrate Reagent A
Nitrate Reagent B
Oxidase Reagent
Potassium hydroxide, 20% with 5% alpha-naphthol
Zinc dust

PREPARED REAGENTS, DISKS AND SLIDES

REAGENT DISKS
Bacitracin (Taxo "A")
Optochin (bile solubility)

ANTIBIOTIC DISKS
Ampicillin, 10 μg
Chloramphenicol, 30 μg
Erythromycin, 15 μg
Gentamycin, 10 μg
Kanamycin, 30 μg
Neomycin, 30 μg
Novobiocin, 30 μg
Penicillin, 10 units
Polymyxin B, 300 units
Strpetomycin, 10 μg
Tetracycline, 30 μg

ANTIBIOTIC SOLUTIONS
Ampicillin
Chloramphenicol
Erythromycin
Gentamicin
Kanamycin
Penicillin
Polymyxin B
Streptomycin
Tetracycline

SEROLOGICAL REAGENTS
Blood Typing Sera
       Anti-A
       Anti-B
       Anti-Rh
Serotyping Reagents
       Polyvalent *Salmonella* O Antiserum Set
       Positive Control Antigen
Febrile Agglutination Set with positive and negative
    controls

MULTIPLE TEST SYSTEM
Enterotube II (Roche Diagnostics)

MISCELLANEOUS SUPPLIES
Disinfectant, laboratory
Ethyl alcohol, 95%
Mineral oil
Sheep blood
Xylol

PREPARED SLIDES
Algae, mixed
Amoeba
Anabaena

Ascaris
Bacteria, 3 types mixed
Blood smear, stained
Capsule stain
Clonorchis
Corynebacterium diphtheriae, metachromatic granule
      stain
Endospore stain
Enterobius
Escherichia coli, Gram stain
Euglena
Flagella stain
Giardia
Gleocapsa
Letter 'e'
Mycobacterium tuberculosis, acid-fast stain
Negative stain
Paramecium
Plasmodium
Pneumocystis carinii
Schistosoma
Staphylococcus aureus, Gram stain
Taenia
Trichinella
Trypanosoma

PRESERVED SPECIMEN
Ascaris
Clonorchis
Enterobius
Schistosoma
Taenia
Trichinella

SUPPLIES
Anaerobic Culturette© Collection and Transport
      System (Becton Dickinson)
Antibiotic disk dispenser
Artificial blood
Bacturcult© Culture Tube (Wampole Laboratories)
Breed slide
Calcium alginate swabs
Calibrated loop
Candle jar
Counting chamber: Neubauer or Petroff-Hauser
Coverslips
Culturette© Collection and Transport System (Becton
      Dickinson)
Gas-Pak anaerobic jar with catalyst, indicator, and
      gas-generator envelopes
Glass spreader
Kimwipes
Lancets
Latex gloves
Membrane filter set-up
Membrane filters, 0.45 μm pore size

Microscope slides
Paper disks, 6.0 mm in diameter
Paraffin block
Pasteur pipettes
Pipetter
Pipettes: serological, 1.0 ml, 5.0 ml, 10.0 ml
Replica plate
Stainless steel pins
Swabs
Toothpicks
Ultraviolet light